Applied Mathematical Sciences
Volume 48

Applied Mathematical Sciences

1. John: Partial Differential Equations, 4th ed. (cloth)
2. Sirovich: Techniques of Asymptotic Analysis.
3. Hale: Theory of Functional Differential Equations, 2nd ed. (cloth)
4. Percus: Combinatorial Methods.
5. von Mises/Friedrichs: Fluid Dynamics.
6. Freiberger/Grenander: A Short Course in Computational Probability and Statistics.
7. Pipkin: Lectures on Viscoelasticity Theory.
8. Giacaglia: Perturbation Methods in Non-Linear Systems.
9. Friedrichs: Spectral Theory of Operators in Hilbert Space.
10. Stroud: Numerical Quadrature and Solution of Ordinary Differential Equations.
11. Wolovich: Linear Multivariable Systems.
12. Berkovitz: Optimal Control Theory.
13. Bluman/Cole: Similarity Methods for Differential Equations.
14. Yoshizawa: Stability Theory and the Existence of Periodic Solutions and Almost Periodic Solutions.
15. Braun: Differential Equations and Their Applications, 3rd ed. (cloth)
16. Lefschetz: Applications of Algebraic Topology.
17. Collatz/Wetterling: Optimization Problems.
18. Grenander: Pattern Synthesis: Lectures in Pattern Theory, Vol I.
19. Marsden/McCracken: The Hopf Bifurcation and its Applications.
20. Driver: Ordinary and Delay Differential Equations.
21. Courant/Friedrichs: Supersonic Flow and Shock Waves. (cloth)
22. Rouche/Habets/Laloy: Stability Theory by Liapunov's Direct Method.
23. Lamperti: Stochastic Processes: A Survey of the Mathematical Theory.
24. Grenander: Pattern Analysis: Lectures in Pattern Theory, Vol. II.
25. Davies: Integral Transforms and Their Applications.
26. Kushner/Clark: Stochastic Approximation Methods for Constrained and Unconstrained Systems.
27. de Boor: A Practical Guide to Splines.
28. Keilson: Markov Chain Models—Rarity and Exponentiality.
29. de Veubeke: A Course in Elasticity.
30. Sniatycki: Geometric Quantization and Quantum Mechanics.
31. Reid: Sturmian Theory for Ordinary Differential Equations.
32. Meis/Markowitz: Numerical Solution of Partial Differential Equations.
33. Grenander: Regular Structures: Lectures in Pattern Theory, Vol. III.
34. Kevorkian/Cole: Perturbation Methods in Applied Mathematics. (cloth)
35. Carr: Applications of Centre Manifold Theory.

(continued after Index)

J. D. Murray

Asymptotic Analysis

With 25 Illustrations

Springer-Verlag
New York Berlin Heidelberg Tokyo

J. D. Murray
Oxford University
Mathematics Institute
24–29 St. Giles'
Oxford OX1 3LB
England

Editors

F. John
Courant Institute of
  Mathematical Sciences
New York University
New York, NY 10012
U.S.A.

J. E. Marsden
Department of
  Mathematics
University of California
Berkeley, CA 94720
U.S.A.

L. Sirovich
Division of
  Applied Mathematics
Brown University
Providence, RI 02912
U.S.A.

AMS Subject Classifications: 40-01, 40A30

Library of Congress Cataloging in Publication Data
Murray, J. D. (James Dickson)
    Asymptotic analysis.
    (Applied mathematical sciences; v. 48)
    Bibliography: p.
    Includes index.
    1. Approximation theory.   2. Asymptotic expansions.
3. Integrals.   4. Differential equations—Numerical
solutions.   I. Title.   II. Series: Applied mathematical
sciences (Springer-Verlag New York Inc.); v. 48.
QA1.A647   vol. 48   510s [515′.24]   83-20426
[QA297.5]

Printed and bound by R. R. Donnelley & Sons, Harrisonburg, Virginia.
Printed in the United States of America.

9 8 7 6 5 4 3 2 1

ISBN 0-387-90937-0 Springer-Verlag New York Berlin Heidelberg Tokyo
ISBN 3-540-90937-0 Springer-Verlag Berlin Heidelberg New York Tokyo

Preface to the Springer Edition

A further practical Chapter 7 has been added on matched asymptotic methods in singular perturbation theory and on multi-scale perturbation methods and suppression of secular terms.

Preface

'A la verité, et ne crains point de l'aduouer, je porterois facilement, au besoing, une chandelle à Saint Michel, l'autre à son serpent.'†

MICHEL DE MONTAIGNE, 1533–1592

THIS book gives an introduction to the most frequently used methods for obtaining analytical approximations to functions defined by integrals or as solutions of ordinary differential equations. The emphasis throughout is on the practical use of the various techniques discussed. Heuristic reasoning, rather than mathematical rigor, is often used to justify a procedure, or some extension of it. This book is mainly intended for mathematicians and scientists whose primary aim is to get answers to practical problems. Frequently rigorous mathematical procedures are not available to deal with many of the problems which arise in practice and one of the aims of this book is to encourage the use of heuristic reasoning to get the solutions. Very often a rigorous development is suggested by a physical one. The philosophy behind the book is that, when solutions to non-standard problems are required, no procedure, be it rigorous or heuristic, should be scorned—a sentiment succinctly summed up by Montaigne.

The first chapter gives an introduction to asymptotic expansions while the next three present the main techniques, with many illustrative examples, for obtaining analytical approximations to integrals.

† This is from Montaigne's essay *De l'utile et de l'honnête* (*Of usefulness and honesty*) and may be freely translated as 'In truth, and I am not afraid to admit it, I would, in need, light a candle to Saint Michael and another to his dragon'.

Chapter 5 deals with the important class of integrals which arise from Fourier and Laplace transform solutions of differential equations. Chapter 6, the last and largest chapter, is concerned with asymptotic methods for ordinary differential equations. This subject is large and very much in current vogue under the title of singular perturbation theory. Here we give some practical methods which are in the nature of an introduction to the subject but which have a surprisingly wide applicability. In the bibliography and throughout the text various books for further reading and reference are listed.

This book is based on lectures given in the mathematics departments at Oxford University and New York University. The material can easily be covered in a single quarter term by final-year undergraduates or first-year graduates with some knowledge of functions of a complex variable and, for the last chapter, of ordinary differential equations.

I would like to thank Dr. A. B. Tayler and Mr. Peter Mitchell for the many helpful suggestions they made during the writing of this book. Finally I would like to express my special indebtedness to Dr. J. R. Ockendon for the many discussions I had with him and for his careful reading of earlier drafts of the text.

Oxford, 1973   J. D. M.

Contents

1. ASYMPTOTIC EXPANSIONS   1

   1.1   Introduction and some of the concepts   1

   1.2   Definitions of asymptotic sequences, expansions, and series   11

2. LAPLACE'S METHOD FOR INTEGRALS   19

   2.1   Integration by parts and Watson's lemma   19

   2.2   Laplace's method   28

3. METHOD OF STEEPEST DESCENTS   40

   3.1   Method of steepest descents   40

   3.2   Illustrative examples   51

4. METHOD OF STATIONARY PHASE   72

   4.1   Method of stationary phase   72

   4.2   Linear dispersive wave motion and the method of stationary phase   79

5. TRANSFORM INTEGRALS   86

   5.1   Transform integrals and their asymptotic evaluation   86

6. DIFFERENTIAL EQUATIONS   99

   6.1   Singularities and asymptotic methods of solution   99

   6.2   Asymptotic solutions with a large or small parameter (WKB method)   111

   6.3   Transition points   131

7. SINGULAR PERTURBATION METHODS   138

   7.1   Basic concepts and introduction to the method of matched expansions   138

   7.2   Method of multiple scales and suppression of secular terms   151

BIBLIOGRAPHY   161

INDEX   163

1

Asymptotic expansions

1.1. Introduction and some of the concepts

EXACT analytical solutions cannot be found for most differential and integral equations which arise in practical situations. By an exact solution we mean one that is given in terms of functions whose properties are known or tabulated: Bessel functions, trigonometric functions, Legendre functions, exponentials, and so on are typical examples. Such a solution may not be particularly useful, however, from either a computational or analytical point of view. For example, a solution which involves a slowly convergent infinite series of Bessel functions is of little use computationally, or even analytically, if we are interested in the dependence of the solution on some parameter of the problem, as is frequently the case. Even within the class of linear equations with linear boundary conditions, to which transform techniques are applicable, the integral representing the inverse of the transform solution may not be integrable in terms of suitable, in the sense of useful, functions. In general asymptotic analysis is that branch of analysis which is concerned with both developing techniques and obtaining approximate analytical solutions to such problems when a parameter or some variable in the equation or integral becomes either large or small or is in the vicinity of a parameter value or point where the solution is not analytic. The ideas developed below are equally applicable under appropriate conditions to differential and integral equations, difference equations, integral evaluation in general, and the evaluation of functions which are represented by series which may be, strictly speaking, divergent, but nevertheless can be used for calculating the function to a high degree of accuracy.

Although some ideas of asymptotics were known in the eighteenth and nineteenth century, it was Poincaré in 1886† who gave a precise definition of what is called an *asymptotic* expansion (see §1.2 below) and laid the foundations of modern asymptotic analysis. Poincaré in 1892‡ developed certain important techniques, on which advances in the past twenty years have essentially been based, in his work on celestial mechanics. The main motivation in the twentieth century has come from the study of fluid mechanics, the governing equations of which are fundamentally nonlinear and display an abundance of parameters which may be small or large with relevant practical situations being covered by both limiting cases. A large and most important development (since 1950) comes under the general heading of *singular perturbation theory,*‖ some aspects of which we discuss in detail in Chapter 6. Again it was within the area of fluid mechanics that the major developments took place. This was partly due to the fact that there was no known exact nontrivial analytical solution of the nonlinear governing equations: the situation is still the same today.

Since many of the most useful techniques in asymptotic analysis are formal or heuristic the trend in recent years has also been to justify and prove the procedures rigorously. There has been also an extension of the areas in which asymptotic techniques have been particularly useful, such as in the traditional physical sciences, astrophysics, the bio-medical sciences, oceanography, traffic studies, epidemic control, population studies, and so on.

Below we shall be concerned with the elements of the subject and the development of the basic ideas and some of the most useful techniques. The emphasis will be on the formal and heuristic development and the applications: a minimum of proofs is given. The books cited more or less cover the spectrum of rigour and applicability as well as discuss some of the less usual aspects of this rather large subject.

As a preliminary, recall the usual *order-notation* denoted by the O- and o-symbols. If $f(z)$ and $g(z)$, two functions of a complex number z, which may be a *parameter* of the problem or an independent variable defined on some domain D, possess limits as $z \to z_0$ in

---

† Henri Poincaré (1886). *Acta Math.* **8**, 295–344.

‡ Henri Poincaré (1892). *Les méthodes nouvelles de la mécanique céleste.* Reprinted Dover Publications Inc., New York (1955).

‖ See, for practical expositions, Van Dyke, M. (1964). *Perturbation methods in fluid mechanics.* Academic Press, New York and Cole, J. D. (1968). *Perturbation methods in applied mathematics.* Blaisdell Publishing Co., Waltham, Mass.

D, then we say that $f(z) = O(g(z))$ as $z \to z_0$ if there exist positive constants K and δ such that $|f| \leq K|g|$ whenever $0 < |z - z_0| < \delta$. If $|f| \leq K|g|$ for all z in D, we say $f(z) = O(g(z))$ in D. If $f(z)$ and $g(z)$ are such that, for any $\varepsilon > 0$, $|f| \leq \varepsilon|g|$ whenever z is in a small δ-neighbourhood of z_0, we say $f(z) = o(g(z))$ as $z \to z_0$. Thus as long as $g(z)$ is not zero in a neighbourhood of z_0, other than possibly at z_0, $f(z) = o(g(z))$ implies that $f/g \to 0$ as $z \to z_0$, while $f(z) = O(g(z))$ implies that f/g is bounded. In the following we shall often take z_0 to be zero or infinity: these occur with most frequency in applications but, in any case, any z_0 can be transformed into either by the change of variable $\xi = z - z_0$ or $\xi = 1/(z - z_0)$.

The O-order is more important than the o-order in asymptotic analysis since it gives more specific knowledge of the function at the point in question. For example, if $f(z) \to 0$ as $z \to 0$ the O-order of $f(z)$ tells us how rapidly $f(z) \to 0$ whereas the o-order merely confirms that $f(z) \to 0$. As a specific example, $\sin z = z + o(z)$ as $z \to 0$ tells us that $\sin z - z \to 0$ faster than z itself. We also have $\sin z = z + o(z^2)$ as $z \to 0$ which tells us that $\sin z - z \to 0$ faster than z^2. However, with $\sin z = z + O(z^3)$ we know that $\sin z - z \to 0$ specifically like z^3. The order function† need not be a simple power: it depends on the refinement that is required or can be obtained. For example, $z^2 \log z + z^3$ is $O(z^2 \log z)$ and $o(z)$ as $z \to 0$ and $O(z^3)$ and $o(z^{n > 3})$ as $z \to \infty$.

The role of the domain D is very important in order relations and in asymptotic analysis in general. For example, $f(z) = 2z + z \cos z = O(z)$ as $z \to \infty$ if z is real since $f(z)$ is bounded by z and $3z$. If z is purely imaginary with $z = iy$, say, then $f(z) = O(ye^y)$ as $z \to \infty$ along the imaginary axis. As another example, consider e^{-z} with the domain D given as (i) $0 < |z| < \infty$, $|\arg z| < \pi/2$ and (ii) $0 < |z| < \infty$, $|\arg z| < \pi$. In (i) with $z = x + iy$, $x > 0$ we have $e^{-z} = e^{-x}e^{-iy} = o(x^n)$ for *all* n as $z \to \infty$ in D. This order relation, namely that a function of z is $o(z^n)$ for *all* n as $z \to \infty$, gives a useful practical criterion for demonstrating the exponential character of a function in situations where little else can be found. In (ii), $x < 0$ is possible in which case we can say nothing about the order relations if we consider only power gauge functions since e^{-x}, with $x < 0$, grows faster than any power of x as $x \to -\infty$. This is then a case where a different order function is required, which here is the obvious one

---

† This is sometimes called the gauge function.

e^{-z} itself, and so $f(z) = O(e^{-z})$. This is also the order function for example, for $ae^{-z} + p(z)$ as $z \to \infty$ in the domain $\pi/2 < \arg z < 3\pi/2$, where a is any constant and $p(z)$ is any polynomial in z.

Various operations hold for order relations, some of which are given as exercises. As an example, suppose $f_n(z) = O(g_n(z))$ for $n = 1, 2, \ldots, N$, then

$$\sum_{n=1}^{N} a_n f_n = O\left(\sum_{n=1}^{N} |a_n| |g_n| \right), \tag{1.1}$$

where the a_n are complex constants. We prove this as follows. Since $f_n = O(g_n)$ we have by definition a real positive constant K_n such that $|f_n| \leq K_n |g_n|$. Let K be the maximum K_n, $n = 1, \ldots, N$, then

$$\left| \sum_{n=1}^{N} a_n f_n \right| \leq \sum_{n=1}^{N} |a_n f_n| \leq \sum_{n=1}^{N} |a_n| |f_n| \leq K \sum_{n=1}^{N} |a_n| |g_n|,$$

which is a statement of (1.1).

Order relations can be integrated but not, in general, differentiated. If a function $f(z, \mu)$ is a function of *two* variables z and μ and $f(z, \mu) = O(g(z, \mu))$ as $z \to z_0$, then frequently, but not always, $\partial f / \partial \mu = O(\partial g / \partial \mu)$ as $z \to z_0$. Although certain general results can be given for differentiation (see the discussion in §1.2 in the case of asymptotic series) in practice each case is best considered separately.

We say that $f(z)$ is *asymptotically equivalent* or *equal* to $g(z)$ under the limit $z \to z_0$ if f and g are such that $\lim_{z \to z_0} f/g = 1$. We write

$$f(z) \sim g(z) \text{ as } z \to z_0 \text{ if } \lim_{z \to z_0} \frac{f(z)}{g(z)} = 1. \tag{1.2}$$

For example, if $f(z) = z^2 + z \log z$, then $f(z) \sim z^2$ as $z \to \infty$ and $f(z) \sim z \log z$ as $z \to 0$. If $f(z) = \cosh z$, then $f(z) \sim 1$ as $z \to 0$ and $f(z) \sim \frac{1}{2} e^z$ as $z \to \infty$.

If $f(z)$ is analytic at some point we have a convergent power series expansion which exhibits in detail the behaviour of the function in the vicinity of that point. It also gives the function to which it is asymptotically equal in the sense of (1.2). So that we can demonstrate some of the differences and similarities of convergent series and asymptotic series which are defined in §1.2, it will be helpful to recall certain properties of convergent series representations of analytic functions. If $f(z)$ is analytic at z_0,

$$f(z) = \sum_{m=0}^{\infty} \alpha_m (z-z_0)^m \text{ for } |z-z_0| < \rho, \qquad (1.3)$$

where the α_m are constants and ρ is the radius of convergence. If z_0 is the point at infinity we have a descending power series

$$f(z) = \sum_{m=0}^{\infty} \frac{a_m}{z^m} \text{ for } |z| > R, \qquad (1.4)$$

where the a_m are constants and R is finite. These power series can be integrated and differentiated term by term for $|z-z_0| < \rho$ in (1.3) and $|z| > R$ in (1.4). Further, if $S_n(z)$ denotes the partial sum to n terms of the series on the right of (1.4), then

$$S_n(z) = \sum_{m=0}^{n} \frac{a_m}{z^m}, \qquad (1.5)$$

and $|f(z)-S_n(z)| \to 0$ not only as $n \to \infty$ for a *fixed* z but *also* as $z \to \infty$ for a *fixed* n. Thus as n increases $S_n(z)$ gives an increasingly more accurate description of $f(z)$. As we shall see below, one very important point about partial sums, which represent the asymptotic expansion of a function, is that they do *not*, in general, have this property but nevertheless give an exceedingly accurate approximation to the function.

If $f(z)$ is not analytic at the point in question it does not possess a convergent Taylor series expansion. In the example above, in which $f(z) = z^2 + z \log z$, $f(z)$ has a branch point at the origin (and infinity) and so does not have a series expansion near $z = 0$ valid uniformly in the neighbourhood of the origin. As $z \to 0$ however, there is no question but that the dominant term is $z \log z$ which gives an increasingly more accurate description of $f(z)$ the smaller z becomes. We shall see below that, even if $f(z)$ is not analytic at the point where we wish to calculate $f(z)$, we can still describe its behaviour by the partial sums $s_n(z)$ of a different kind of series approximation called an *asymptotic expansion*. Such series are usually not convergent but nevertheless $|f(z)-s_n(z)| \to 0$ for a *fixed* n as z tends to the point in question, z_0 say. It will be seen below that the domain of z as $z \to z_0$ is crucially important, just as it was, for example, in determining the order relation as $z \to \infty$ in the above case where $f(z) = 2z + z \cos z$.

The following example demonstrates some of the differences between convergent and asymptotic series and introduces some of the concepts of asymptotic expansions discussed formally in §1.2 below. It is also a good example of the procedure described in §2.1.

Consider the real exponential function Ei(x), where x is real and positive, defined by

$$\mathrm{Ei}(x) = \int_{x}^{\infty} e^{-t} t^{-1} \, dt \tag{1.6}$$

and let us look for an analytical approximation to Ei(x) for x large and positive. Integration by parts gives

$$\mathrm{Ei}(x) = \left[-\frac{e^{-t}}{t} \right]_{x}^{\infty} - \int_{x}^{\infty} e^{-t} t^{-2} \, dt$$

$$= \frac{e^{-x}}{x} + \left[\frac{e^{-t}}{t^2} \right]_{x}^{\infty} + 2 \int_{x}^{\infty} e^{-t} t^{-3} \, dt,$$

and on repeated integration by parts we have

$$\mathrm{Ei}(x) = e^{-x} \left\{ \frac{1}{x} - \frac{1}{x^2} + \frac{2!}{x^3} - \frac{3!}{x^4} + \ldots + \frac{(-1)^{n+1}(n-1)!}{x^n} \right\}$$

$$+ (-1)^n n! \int_{x}^{\infty} e^{-t} t^{-(n+1)} \, dt$$

$$= s_n(x) + r_n(x), \tag{1.7}$$

where the partial sum $s_n(x)$ and the remainder $r_n(x)$ are given by

$$s_n(x) = e^{-x} \left\{ \frac{1}{x} - \frac{1}{x^2} + \frac{2!}{x^3} - \frac{3!}{x^4} + \ldots + \frac{(-1)^{n+1}(n-1)!}{x^n} \right\},$$

$$r_n(x) = (-1)^n n! \int_{x}^{\infty} e^{-t} t^{-(n+1)} \, dt. \tag{1.8}$$

The series for which $s_n(x)$ is the partial sum is divergent for any fixed x since the nth term tends to infinity as $n \to \infty$. Of course $r_n(x)$ is also unbounded as $n \to \infty$ since $s_n(x) + r_n(x)$ must be bounded because Ei(x) for $x > 0$ is bounded: this is also clear from the definition of $r_n(x)$ in (1.8). Suppose, however, we consider n *fixed* and let x become large then, from (1.8),

$$|r_n(x)| = n! \int_x^{\infty} e^{-t} t^{-(n+1)} \, dt < \frac{n!}{x^{n+1}} \int_x^{\infty} e^{-t} \, dt$$

$$= \frac{n! \, e^{-x}}{x^{n+1}} \to 0 \text{ as } x \to \infty. \qquad (1.9)$$

Here $n! \, e^{-x} x^{-(n+1)}$ is the magnitude of the next $((n+1)$th) term in the series for $s_n(x)$. Thus for n *fixed* the ratio of $r_n(x)$ to the last term in $s_n(x)$ is such that

$$\left| r_n(x)/(n-1)! \, x^{-n} \, e^{-x} \right| < \frac{n}{x} \to 0 \text{ as } x \to \infty, \, n \text{ fixed.} \qquad (1.10)$$

Thus if x is sufficiently large and n *finite*, $s_n(x)$ gives a good approximation to Ei(x): the accuracy of the approximation increases with increasing x for n fixed, as is clear from (1.10). By a good approximation we mean that the error is a small fraction of the exact value. The error here in approximating Ei(x) by $s_n(x)$ is of the order of the first term neglected in $s_n(x)$, namely, $(-1)^n \, n! \, e^{-x} x^{-(n+1)}$. This criterion is not uniformly applicable for this approach, but it is at least the case when the remainder $r_n(x)$ alternates in sign with n: this is reminiscent of the situation which obtains with convergent oscillating series.

From the definition (1.2), Ei(x) is asymptotically equal or equivalent to $s_n(x)$ in (1.8) for all fixed n as may be checked, and, since n is at our disposal, we write

$$\text{Ei}(x) \sim e^{-x} \left(\frac{1}{x} - \frac{1}{x^2} + \frac{2!}{x^3} + \ldots \right) \text{ as } x \to \infty, \qquad (1.11)$$

the right-hand side of which is the *asymptotic expansion* of Ei(x) as $x \to \infty$.

The question now naturally arises as to how we choose the optimal n so that $s_n(x)$ gives a best approximation to Ei(x) for a given x. Clearly for x sufficiently large the terms in $s_n(x)$ from (1.8) will successively decrease initially, for example $2! x^{-3} < x^{-2}$ for x large enough. However, at some value of n, N say, the terms with $n > N$ will start to increase successively for a given x, however large, since the nth term $(-1)^{n+1} (n-1)! \, e^{-x} x^{-n}$ is unbounded as $n \to \infty$. In the class of problems of which this is an example, the error in stopping $s_n(x)$ at a given n is of the order of the first term neglected, and so the optimal place to terminate $s_n(x)$ is with $n = N$: of course, N is a function of x. Figure 1.1 illustrates, in a typical asymptotic situation, the magnitude of the terms in the series as a function of their order.

The practical procedure for calculating $\mathrm{Ei}(x)$ for a given fixed x is to evaluate the successive terms in $s_n(x)$ and to terminate the series when the first term found to be greater than the previous one is obtained. Here we look for the n such that the modulus of the ratio

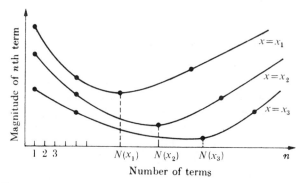

Fig. 1.1. Optional number of terms N for different x: $x_1 < x_2 < x_3$.

of the $(n+1)$th to the nth term in $s_n(x)$ is as close as it can get to (but less than) unity. From (1.8) this ratio is

$$n!\,e^{-x}\,x^{-(n+1)}/(n-1)!\,e^{-x}\,x^{-n} = nx^{-1},$$

and so N must be taken to be the largest integral part of the given x.

Convergence of a series is not necessarily of practical value from a computational point of view unless, of course, it is rapidly convergent, since convergence depends on the nth term for indefinitely large n. What is most desirable is an approximation which requires only a few terms. Asymptotic approximations which give *divergent* series, as in the example above, are usually of considerably *more* practical use than if a convergent series is a result of the exercise. (An example where a convergent series is not of much practical use from a computational point of view is the series for the exponential e^x for large x.) Such asymptotic approximations can be remarkably accurate. For example, for $x = 10$, even though the optimal N is thus 10, $s_4(10)$ gives $\mathrm{Ei}(10)$ with an error of less than 0·003 per cent.

Suppose now we choose x in (1.6) to be complex, equal to z, say, and consider

$$\mathrm{Ei}(z) = \int_z^\infty e^{-t} t^{-1} \, dt,$$

where we take $\mathrm{Im}\, z \neq 0$ and, by way of example, the path of integration from $\mathrm{Rl}\, z \geq 0$ to infinity along a line parallel to the real axis. Integrating by parts in exactly the same way as before we get

$$s_n(z) = e^{-z} \left\{ \frac{1}{z} - \frac{1}{z^2} + \frac{2!}{z^3} + \ldots + \frac{(-1)^{n+1}(n-1)!}{z^n} \right\},$$

$$r_n(z) = (-1)^n n! \int_z^\infty e^{-t} t^{-(n+1)} \, dt,$$

and here with the path of integration as given, with $t = \rho + i\sigma$, say, where $\sigma = \mathrm{Im}\, z$ is constant, $dt = d\rho$, $x \leq \rho \leq \infty$ and so

$$|r_n(z)| \leq n! \int_x^\infty \left| e^{-\rho - i\sigma} (\rho + i\sigma)^{-(n+1)} \right| d\rho$$

$$= n! \int_x^\infty e^{-\rho} |\rho + i\sigma|^{-(n+1)} \, d\rho$$

$$< n! \, |z|^{-(n+1)} \int_x^\infty e^{-\rho} \, d\rho$$

$$= n! \, |z|^{-(n+1)} e^{-x}$$

$$= n! \left| e^{-z} z^{-(n+1)} \right|,$$

which is again simply the modulus of the next term in the series $s_n(z)$ and so the error in terminating the series approximation $s_n(z)$ for $\mathrm{Ei}(z)$ is again of the order of the first term neglected.

Exercises

1. Show that the following order relations hold and give, where appropriate, any restrictions on the domain of z:

   (i) $\left. \begin{array}{l} 1 - \cos^2 z = O(z^2) \\ = o(z) \end{array} \right\}$ as $z \to 0$;

   (ii) $\log(1 + z) = O(z)$ as $z \to 0$;

   (iii) $(\log z)^2 = o(z^{1/3})$ as $z \to \infty$;

   (iv) $\left. \begin{array}{l} z + \log(a + z) = O(1) \\ = \log a + O(z) \end{array} \right\}$ as $z \to 0$ with $a \neq 1$;

(v) $\dfrac{z^2}{1+z^3}+\log (1+z^2) = o(z)$ $\left.\begin{array}{l} \\ = O(z^2) \end{array}\right\}$ as $z \to 0$;

$\qquad\qquad\qquad\qquad = O(\log z)$ as $z \to \infty$;

(vi) $a^z + z = O(1)$ as $z \to 0$ with a complex;

(vii) $e^{-1/z} = o(z^n)$ for *all* n as $z \to 0$;

$\qquad\qquad = O(1)$ as $z \to \infty$;

(viii) $\sec^{-1}(1+z) = O(z^{\frac{1}{2}}) = o(1)$ as $z \to 0$;

(ix) $\sinh 1/z = O(e^{1/z})$ as $z \to 0$;

$\qquad\qquad = o(1)$ as $z \to \infty$;

(x) $\operatorname{sech}^{-1} z = O(\log z)$ as $z \to 0$.

2. Are there any domains such that $e^{-z} = O(z^{\alpha})$ for any α?

3. If $f(z) = O(z)$ and $g(z) = O(z^3)$ as $z \to 0$, and $f(z) = O(z)$ and $g(z) = O(z^2)$ as $z \to \infty$, show that
$$a f(z) + b g(z) = O(f(z)) \quad \text{as } z \to 0 \left.\begin{array}{l} \\ = O(g(z)) \quad \text{as } z \to \infty \end{array}\right\}$$
for non-zero constants a and b.

4. If $f = O(g)$ show that

(i) $O(o(f)) = o(O(f)) = o(g)$;

(ii) $O(f)O(g) = O(fg)$;

(iii) $O(f)o(g) = o(f)o(g) = o(fg)$.

5. Show that if

(i) $f = O(g)$, then $|f|^{\alpha} = O(|g|^{\alpha})$ for $\alpha > 0$;

(ii) $f_i = O(g_i), i = 1, 2, \ldots, n$ and $|g_i| \le |g|$, then $\displaystyle\sum_{i=1}^{n} a_i f_i = O(g)$,

where a_i, $i = 1, 2, \ldots, n$ are constants;

(iii) $f_i = O(g_i)$, $i = 1, 2, \ldots, n$, then $\displaystyle\prod_{i=1}^{n} f_i = O(\prod_{i=1}^{n} g_i)$;

(iv) in (i), (ii), and (iii) O is replaced by o, the relations still hold.

6. Show that if

(i) $f(z) = z + z \sin z, \displaystyle\int_{0}^{z} f(z)\, dz = O(\int_{0}^{z} (O(f(z))$ as $z \to 0)\, dz)$;

(ii) $f(z) = z + z \log (1+z), \displaystyle\int_{0}^{z} f(z)\, dz = O(z^2 \log z)$ as $z \to \infty$.

7. Find the asymptotic expansion of $\operatorname{Ei}_m(x) = \displaystyle\int_{x}^{\infty} e^{-t} t^{-m}\, dt$, where m is a positive integer and x is real, positive, and large.

8. Find the asymptotic expansion of $I(x) = \int_x^\infty e^{it} t^{-1} \, dt$ for x real, positive, and large and hence or otherwise show that the cosine integral

$$Ci(x) = Rl \, I(x) = \int_x^\infty \frac{\cos t}{t} \, dt$$

$$\sim \left(\frac{1}{x^2} - \frac{3!}{x^4} + \ldots \right) \cos x - \left(\frac{1}{x} - \frac{2!}{x^3} + \ldots \right) \sin x.$$

9. Show that the Laplace transform of $(1+t^2)^{-1}$, namely

$$\int_0^\infty e^{-pt} (1+t^2)^{-1} \, dt \sim \frac{1}{p} - \frac{2!}{p^3} + \ldots + \frac{(-1)^n(2n)!}{p^{2n+1}} + \ldots$$

as $p \to \infty$ in the domain $|\arg p| < \pi/2$, by considering

$$(1+t^2)^{-1} = 1 - t^2 + t^4 + \ldots + (-1)^{n-1} t^{2n-2} + \frac{(-1)^n t^{2n}}{1+t^2}$$

in the integral.

1.2. Definitions of asymptotic sequences, expansions, and series

A finite or infinite sequence of functions $\{\phi_n(z)\}$, $n = 1, 2, \ldots$ is an *asymptotic sequence* as $z \to z_0$ if, for *all* n,

$$\phi_{n+1}(z) = o(\phi_n(z)) \text{ as } z \to z_0, \tag{1.12}$$

that is, $\lim_{z \to z_0} \phi_{n+1}/\phi_n = 0$. Definition (1.12) is on the understanding that there are no zeros of $\phi_n(z)$, other than possibly z_0, in a neighbourhood of z_0. Some examples of asymptotic sequences, where $n = 0, 1, 2, \ldots$, are

$\{(z - z_0)^n\}$, $z \to z_0$,

$\{(\log z)^{-n}\}$, $z \to \infty$,

$\{e^z z^{-a_n}\}$, $z \to \infty$, a_n real with $a_{n+1} > a_n$,

$1, z, z^2 \log z, z^2, z^3 (\log z)^2, z^3 \log z, z^3, z^4 (\log z)^3, \ldots, z \to 0$.

That these are asymptotic sequences can be verified immediately by using (1.12). As an example, consider the third of these:

$$\frac{\phi_{n+1}(z)}{\phi_n(z)} = \frac{e^z z^{-a_{n+1}}}{e^z z^{-a_n}} = z^{-(a_{n+1} - a_n)} \to 0 \text{ as } z \to \infty$$

since $a_{n+1} - a_n > 0$.

Using the various operations for order relations in §1.1 and the exercises above, together with (1.12), new asymptotic sequences can be formed by appropriately combining asymptotic sequences. Again, integration of an asymptotic sequence gives an asymptotic sequence but differentiation does not in general do so.

If $\{\phi_n(z)\}$ is an asymptotic sequence of functions as $z \to z_0$, we say that $\sum_{n=1} a_n \phi_n(z)$, where the a_n are constants (with the upper limit omitted), is an *asymptotic expansion* or *asymptotic approximation* of the function $f(z)$ if for each N

$$f(z) = \sum_{n=1}^{N} a_n \phi_n(z) + o(\phi_N(z)) \text{ as } z \to z_0. \qquad (1.13)$$

This is essentially Poincaré's (1886) definition of an asymptotic expansion. If the sequence $\{\phi_n(z)\}$ is asymptotic in some restricted domain, the asymptotic expansion can only be valid in the same domain. Note that (1.13) may be written as

$$f(z) = \sum_{n=1}^{N-1} a_n \phi_n(z) + O(\phi_N(z)), \qquad (1.14)$$

which implies that the error is of the same order of magnitude as the first term omitted. We emphasize the freedom to choose N as we like by using the asymptotic equivalence definition (1.2) by writing

$$f(z) \sim \sum_{n=1}^{\infty} a_n \phi_n(z) \text{ as } z \to z_0. \qquad (1.15)$$

The expansion for $\text{Ei}(x)$ in (1.11) is such an expansion. Note that (1.15) does not imply in any sense that $\sum_{n=1}^{\infty} a_n \phi_n(z)$ exists as a convergent series; it merely implies (1.13) for all given N. An asymptotic expansion may, of course, be convergent but, as discussed above, if such is the case, it is usually less useful than if it is divergent. This is so since, for any z, only a few terms are required in the case of a divergent asymptotic series to give a very accurate approximation to the function. If $f(z)$ is analytic at z_0 then its Taylor series about z_0 is a convergent asymptotic expansion (exercise 3).

If an asymptotic expansion of $f(z)$, such as (1.13), exists for a *given* asymptotic sequence $\{\phi_n(z)\}$, it is *unique*; the a_n are uniquely determined by successively applying the following limits:

$$a_1 = \lim_{z \to z_0} \frac{f(z)}{\phi_1(z)},$$

$$a_2 = \lim_{z \to z_0} \frac{f(z) - a_1 \phi_1(z)}{\phi_2(z)},$$

$$\text{---} \quad \text{---} \quad \text{---} \quad \text{---} \quad \text{---} \quad \text{---} \quad (1.16)$$

$$a_N = \lim_{z \to z_0} \left\{ \frac{f(z) - \sum_{n=1}^{N-1} a_n \phi_n(z)}{\phi_N(z)} \right\}.$$

The first non-zero term in the asymptotic expansion $\sum\limits_{n=1}^{\infty} a_n \phi_n(z)$ is called the leading or dominant term in the expansion; and we frequently write, if $a_1 \neq 0$ for example, $f(z) \sim a_1 \phi_1(z)$, as $z \to z_0$ in the appropriate domain, and this has the meaning implied by the asymptotic equivalence relation (1.2). In many practical examples the appropriate asymptotic sequence may not be known and the leading term in an expansion is all that can be found, and frequently is all that is required.

The commonest asymptotic sequence which occurs is the power sequence $\{(z - z_0)^n\}$ as $z \to z_0$ in some domain. The point z_0 may, without loss of generality, be taken to be the point at infinity or the origin by the appropriate change of variable mentioned in §1.1, namely $\xi = 1/(z - z_0)$ or $\xi = z - z_0$. We shall consider, therefore, z_0 to be the point at infinity. A typical expansion is then of the form

$$f(z) \sim g(z) \sum_{n=1}^{\infty} \frac{a_n}{z^n} \text{ as } z \to \infty, \qquad (1.17)$$

where $g(z)$ is some function, for example e^{-z} in the asymptotic expansion given for $\mathrm{Ei}(z)$ in §1.1. Asymptotic expansions based on asymptotic power sequences are often called *asymptotic power series*. The asymptotic expansion given for $\mathrm{Ei}(z)$ in §1.1 is an asymptotic power series in the sequence $\{(-1)^{n+1} (n-1)! \, z^{-n}\}$ as $z \to \infty$ in the domain $|\arg z| < \pi/2$.

The asymptotic expansion of a function $f(z)$ depends on the specific sequence $\{\phi_n(z)\}$ chosen, and so a function may possess several asymptotic expansions, which can be very useful when expansions are required in different domains, as is often the case. This point is simply illustrated by considering power series expansions for $f(z) = 1/(z - 1)$ for $|z| > 1$. We have

$$\frac{1}{z - 1} \sim \sum_{n=1}^{\infty} \frac{1}{z^n} \text{ as } z \to \infty, \qquad (1.18)$$

and since, for example,

$$\frac{1}{z^2-1} \sim \sum_{n=1}^{\infty} \frac{1}{z^{2n}} \text{ as } z \to \infty,$$

we also have

$$\frac{1}{z-1} \sim (z+1) \sum_{n=1}^{\infty} \frac{1}{z^{2n}} \text{ as } z \to \infty. \tag{1.19}$$

Expansions (1.18) and (1.19) are different convergent asymptotic power series for the same function based on the two sequences $\{z^{-n}\}$ and $\{z^{-2n}\}$ respectively.

Not only may a function possess more than one asymptotic expansion, a given asymptotic expansion may be the expansion for more than one function. Consider first e^{-z} in terms of the sequence $\{z^{-n}\}$ as $z \to \infty$ and write

$$e^{-z} \sim \sum_{n=0}^{\infty} \frac{\alpha_n}{z^n} \text{ as } z \to \infty \text{ in the domain } |\arg z| < \frac{\pi}{2}.$$

Using (1.16), however, $\alpha_0 = \lim_{z \to \infty} e^{-z} = 0, \alpha_1 = \lim_{z \to \infty} z e^{-z} = 0, \ldots$ $\alpha_n = \lim_{z \to \infty} z^n e^{-z} = 0$ for *all* n. Thus if $f(z)$ possesses an asymptotic power series expansion $\sum_{n=0}^{\infty} a_n/z^n$ as $z \to \infty$ in the domain $|\arg z| < \pi/2$, then $\sum_{n=0}^{\infty} a_n/z^n$ is also the asymptotic power series of $f(z)+e^{-z}$ as $z \to \infty$ in the domain $|\arg z| < \pi/2$.

Various operations hold for asymptotic expansions. For example, if $f(z) \sim \sum_{n=1} a_n \phi_n(z)$ and $g(z) \sim \sum_{n=1} b_n \phi_n(z)$ in D, then with α and β complex constants,

$$\alpha f(z)+\beta g(z) \sim \sum_{n=1} (\alpha a_n+\beta b_n) \phi_n(z) \text{ in } D. \tag{1.20}$$

The proof follows immediately using the definition (1.13) and order relation operations (exercise 5(i)).

The product of $f(z)$ and $g(z)$ may not necessarily have an asymptotic expansion. Formal multiplication of the two asymptotic expansions gives a system of terms $\phi_n(z) \phi_m(z)$, $n = 1, 2, \ldots$ and $m = 1, 2, \ldots$ which may not necessarily be able to be arranged as an asymptotic sequence. In the case of asymptotic power series, however, the product may be an asymptotic expansion, one of the crucial factors in this case being the domains of the expansions for $f(z)$ and

$g(z)$. If the domain is the same for each and

$$f(z) = \sum_{n=0}^{N} \frac{a_n}{z^n} + O(z^{-(N+1)})$$

$$g(z) = \sum_{n=0}^{N} \frac{b_n}{z^n} + O(z^{-(N+1)})$$

$\left.\begin{array}{c} \\ \\ \end{array}\right\}$ as $z \to \infty$ in D,   (1.21)

then by simply multiplying the two right-hand sides of (1.21) we get (exercise 5(ii))

$$f(z)\, g(z) = \sum_{n=0}^{N} \frac{c_n}{z^n} + O(z^{-(N+1)}) \text{ as } z \to \infty \text{ in } D$$

where

$$c_n = a_0\, b_n + a_1\, b_{n-1} + \ldots + a_{n-1}\, b_1 + a_n\, b_0.$$

$\left.\begin{array}{c} \\ \\ \\ \end{array}\right\}$ (1.22)

From (1.22) it follows that $\{f(z)\}^m$, where m is a positive integer, possesses an asymptotic power series as $z \to \infty$ in D, and hence that any polynomial function of $f(z)$ and any rational function of $f(z)$, where the denominator is never zero, also do so.

Asymptotic power series can be integrated term by term and the result is an asymptotic power series. Suppose

$$f(z) = \sum_{n=0}^{N} \frac{a_n}{z^n} + O(z^{-(N+1)}) \text{ as } z \to \infty \text{ in } D,$$

then

$$f(z) - a_0 - \frac{a_1}{z} = \sum_{n=2}^{N} \frac{a_n}{z^n} + r(z),$$

where $r(z) = O(z^{-(N+1)})$ as $z \to \infty$, and so

$$\int_z^\infty \left\{ f(\xi) - a_0 - \frac{a_1}{\xi} \right\} \mathrm{d}\xi = \sum_{n=2}^{N} \frac{a_n}{(n-1)\, z^{n-1}} + \int_z^\infty r(\xi)\, \mathrm{d}\xi.$$

Now $|r(z)| \leq K|z^{-(N+1)}|$ and if we take, for example, the path of integration along a path in D with fixed argument, then

$$\left| \int_z^\infty r(\xi)\, \mathrm{d}\xi \right| \leq K \int_{|z|}^\infty |\xi|^{-(N+1)}\, \mathrm{d}|\xi| = \frac{K}{N} |z|^{-N},$$

and we have

$$\int_z^\infty \left\{ f(\xi) - a_0 - \frac{a_1}{\xi} \right\} \mathrm{d}\xi = \sum_{n=2}^{N} \frac{a_n}{(n-1)\, z^{n-1}} + o(z^{-N+1}).$$

Since the above holds for all N,

$$\int_z^\infty \left\{ f(\xi) - a_0 - \frac{a_1}{\xi} \right\} d\xi \sim \sum_{n=1}^N \frac{a_{n+1}}{nz^n} \text{ as } z \to \infty \text{ in } D. \quad (1.23)$$

The result (1.23) holds for all paths of integration in D.

Term-by-term differentiation of an asymptotic expansion does not necessarily give an asymptotic expansion. However, if $f(z)$ and $f'(z)$ both possess asymptotic power series expansions as $z \to \infty$ in D, then

$$\left. \begin{array}{l} f(z) \sim \displaystyle\sum_{n=0}^\infty \frac{a_n}{z^n} \\[2em] f'(z) \sim - \displaystyle\sum_{n=1}^\infty \frac{na_n}{z^{n+1}} \end{array} \right\} \text{ as } z \to \infty \text{ in } D. \quad (1.24)$$

To prove this we simply use the result (1.23) above with $f'(\xi)$ and its asymptotic power series in place of $f(\xi)$ and its series expansion; (1.24) is obtained immediately.

If $f(z)$ is single-valued and analytic for all $|z| > R$ it possesses a convergent power series expansion valid for *all* arg z. By the uniqueness property of asymptotic series, this convergent power series must be the same as the asymptotic series for $f(z)$ as $z \to \infty$ based on $\{z^{-n}\}$ and so this asymptotic expansion is clearly differentiable, and integrable, for all $|z| > R$. If $f(z)$ is *not* analytic everywhere it *cannot* have a single asymptotic power series expansion in $\{z^{-n}\}$ valid for all arg z. As mentioned above, different expansions may be necessary for different argument domains. Various examples are given in subsequent sections.

Asymptotic problems arise in a variety of classical problems which can be treated by classical means such as the inversion of implicit functions by Lagrange's method†; iterative procedures, with applications to difference equations, for finding roots of equations by Newton's method; the Euler–Maclaurin procedure†, which is of current use and interest, for summing series and so on. Such problems will not be discussed here: the book by de Bruijn (1958) covers these topics. The emphasis below is on the more modern aspects of the subject and the techniques which have the widest applicability. The techniques for integrals are of particular importance since they represent, in a large number of situations, solutions of differential

---

† See, for example, Whittaker, E. T. *and* Watson, G. N. (1946). *Modern analysis.* Cambridge University Press.

equations which may, for example, have arisen from transform methods. Integrals are also frequently used to define special functions such as Bessel functions, the Gamma function, Legendre functions, and so on and are thus of particular importance in analytic function theory. The large or small variable in the asymptotic analysis may be an independent variable or a parameter in the differential equation, for example, which gave rise to the integral.

Exercises

1. Verify that the following are asymptotic sequences and give in both cases the domain of z:

   (i) $\{\log(1+z^n)\}$, $n = 0, 1, ..., z \rightarrow 0$;

   (ii) $\{e^{-nz}z^{-a_n}\}$, $n = 0, 1, ..., z \rightarrow \infty$, $\text{Rl}\, a_{n+1} > \text{Rl}\, a_n, |\arg z| \leq \pi/2$.

2. Show that $[x^n\{a+\cos(x^{-n})\}]$, $n = 0, 1, ...$, with x real and $a > 1$ is an asymptotic sequence as $x \rightarrow 0$ but, with the same conditions,

$$\left[\frac{d}{dx} x^n\{a+\cos(x^{-n})\} \right]$$

is not an asymptotic sequence.

3. Show that the Taylor series of a function about a point z_0 where it is analytic is an asymptotic expansion.

4. Find two asymptotic power series for $1/(z+1)$ as $z \rightarrow \infty$ based on the two asymptotic sequences
$$\{z^{-2n}\} \text{ and } \{(-1)^n z^{-3n}\}, \; n = 0, 1, \ldots .$$

5. Prove that if

   (i) $f(z) \sim \displaystyle\sum_{n=1} a_n \phi_n(z)$ and $g(z) \sim \displaystyle\sum_{n=1} b_n \phi_n(z)$ in a domain D, then

   for any constants α and β
   $$\alpha f(z) + \beta g(z) \sim \sum_{n=1} (\alpha a_n + \beta b_n) \phi_n(z);$$

   (ii) $f(z) = \displaystyle\sum_{n=0}^{N} \frac{a_n}{z^n} + O(z^{-(N+1)})$

   $\left.\begin{array}{l} \\ \\ \end{array}\right\}$ as $z \rightarrow \infty$ in D,

   $g(z) = \displaystyle\sum_{n=0}^{N} \frac{b_n}{z^n} + O(z^{-(N+1)})$

   then

   $$f(z)g(z) = \sum_{n=0}^{N} \frac{c_n}{z^n} + O(z^{-(N+1)}) \text{ as } z \rightarrow \infty \text{ in } D$$

   where

   $$c_n = a_0 b_n + a_1 b_{n-1} + \ldots + a_{n-1} b_1 + a_n b_0.$$

6. If $f(z) = \sum_{n=0}^{N} \dfrac{a_n}{z^n} + O(z^{-(N+1)})$ in the domain $\alpha_1 < \arg z < \alpha_2$

and

$g(z) = \sum_{n=0}^{N} \dfrac{b_n}{z^n} + O(z^{-(N+1)})$ in the domain $\beta_1 < \arg z < \beta_2$, as $z \to \infty$,

discuss the possible asymptotic power series expansion in $\{z^{-n}\}$ of $f(z)g(z)$ as $z \to \infty$.

7. If $f(z) \sim \sum_{n=0}^{\infty} \dfrac{a_n}{z^n}$ as $z \to \infty$ and $a_0 \neq 0$, find the first few terms in an asymptotic power series in $\{z^{-n}\}$ as $z \to \infty$ of

  (i) $1/f(z)$, (ii) $g(f(z))$ where $g(f)$ is a polynomial in f.

8. If $f(z) \sim \sum_{n=0}^{\infty} a_n z^n$ and $f'(z)$ possesses an asymptotic power series expansion as $z \to 0$ in D, prove that

  (i) $\int_{0}^{z} f(\xi)d\xi \sim \sum_{n=0}^{\infty} \dfrac{a_n}{n+1} z^{n+1}$ as $z \to 0$ in D;

  (ii) $f'(z) \sim \sum_{n=1}^{\infty} na_n z^{n-1}$ as $z \to 0$ in D.

2

Laplace's method for integrals

2.1. Integration by parts and Watson's lemma

We have already seen in §1.1 that integration by parts is one way of finding asymptotic approximations to integrals. It is one of the simplest procedures but it is rather limited in its applicability. The procedure is essentially to integrate by parts and then show that the resulting series is asymptotic by estimating the remainder which is in the form of an integral: this is exactly what was done in §1.1 to obtain (1.11) for Ei(x) as $x \to \infty$. We include in this procedure the technique where the integrand is expanded as a series and the asymptotic series is obtained by integrating term by term: the asymptotic nature of the resulting series again depends on the estimation of an integral remainder (see, for example, §1.1 exercise 9). A survey of these methods, presented by way of specific examples, is given in the book by Copson (1965).

One important point which should always be kept in mind, particularly with these methods, is that if a convergent asymptotic series is a result of the exercise, it is of less practical use than if a divergent asymptotic approximation results. To emphasize this, and to demonstrate what can be done in many cases, consider the incomplete Gamma function defined by

$$\gamma(a, x) = \int_0^x e^{-t} t^{a-1} \, dt, \tag{2.1}$$

where we shall take x and a to be real and positive, with $a > 0$ to ensure convergence of the integral at $t = 0$. The complete Gamma function $\Gamma(a)$ for real $a > 0$ is defined by $\gamma(a, \infty)$ and so

$$\Gamma(a) = \int_0^\infty e^{-t} t^{a-1} \, dt, \, a > 0. \tag{2.2}$$

Recalling that $\Gamma(a + 1) = a\Gamma(a)$ and so on, we have, in the case where a is a positive integer m, say, $\Gamma(m + 1) = m!$ When $a = \frac{1}{2}$, (2.2) is an error function and $\Gamma(\frac{1}{2}) = \sqrt{\pi}$.

For $x \to 0$ we might expand the integrand in (2.1) in powers of t and integrate term by term to get

$$\gamma(a, x) = \int_0^x \left(1 - t + \frac{t^2}{2!} - \frac{t^3}{3!} + \ldots\right) t^{a-1} \, dt$$

$$= x^a \sum_{n=0}^{\infty} \frac{(-1)^n \, x^n}{(a+n) \, n!} \tag{2.3}$$

But (2.3) converges for *all* x and so it could be used to calculate $\gamma(a, x)$ for all x. For all but small x, however, the series (2.3) is of little practical use, and in fact becomes progressively less useful the larger x becomes. For large x we can obtain a different asymptotic series, which is divergent, by writing

$$\gamma(a, x) = \int_0^{\infty} e^{-t} t^{a-1} \, dt - \int_x^{\infty} e^{-t} t^{a-1} \, dt$$

$$= \Gamma(a) - \mathrm{Ei}_{1-a}(x), \tag{2.4}$$

where $\Gamma(a)$ is the Gamma function (2.2) and $\mathrm{Ei}_{1-a}(x)$ is the exponential function defined by the second integral in (2.4): the case $a = 0$ was discussed in detail in §1.1 (see also §1.1 exercise 7). Successive integration by parts of $\int_x^{\infty} e^{-t} t^{a-1} \, dt$ in a similar way to that used for $\mathrm{Ei}(x)$ in §1.1 now gives a divergent asymptotic series. Here,

$$\mathrm{Ei}_{1-a}(x) = \int_x^{\infty} e^{-t} t^{a-1} \, dt$$

$$= e^{-x} x^{a-1} + (a-1) \int_x^{\infty} e^{-t} t^{a-2} \, dt,$$

which, on repeated integration by parts, gives

$$\mathrm{Ei}_{1-a}(x) = e^{-x} \left\{ x^{a-1} + (a-1) x^{a-2} + \ldots + \right.$$
$$+ (a-1)(a-2) \ldots (a-N+1) x^{a-N} \} +$$
$$+ (a-1)(a-2) \ldots (a-N) \int_x^{\infty} e^{-t} t^{a-N-1} \, dt.$$

For a fixed N,

$$\left| \int_x^\infty e^{-t} t^{a-N-1} \, dt \right| < x^{a-N-1} \int_x^\infty e^{-t} \, dt, \text{ for } N > a-1,$$

$$= x^{a-N-1} e^{-x}$$

$$= o(x^{a-N} e^{-x}) \text{ as } x \to \infty;$$

and so the above expansion for $Ei_{1-a}(x)$ is asymptotic. Using this, we finally get an asymptotic expansion for $\gamma(a, x)$ from (2.4) as

$$\gamma(a, x) \sim \Gamma(a) - e^{-x} x^a \left\{ \frac{1}{x} + \frac{a-1}{x^2} + \frac{(a-1)(a-2)}{x^3} + \cdots \right\}.$$

$$(2.5)$$

The asymptotic series (2.5) is divergent. By taking the ratio of the $(N+1)$th to the Nth term in (2.5), in the same way as was done in §1.1 to obtain the optimal N for a given x, we here get N given by $|a-N|$ equal to the largest integral part of x. If a is an integer the series terminates: in this case the right-hand side of (2.5) is exactly, rather than asymptotically, equal to $\gamma(a, x)$.

At this stage we might ask for the asymptotic expansion of $\Gamma(a)$ when a is large. Straightforward integration by parts is not applicable and a different approach is required: this problem is discussed below in §2.2.

One important class of integrals which, under certain conditions, is amenable to this method of integration by parts is the class of *Laplace integrals* of the form

$$f(x) = \int_0^\infty e^{-xt} \phi(t) \, dt,$$

$$(2.6)$$

where the allowable $\phi(t)$ are those which make the integral exist for the given x. When x is complex, such integrals represent the Laplace transform of the function $\phi(t)$ and are widely used, of course, in solving certain linear differential and integral equations.

We now prove *Watson's lemma*† which provides asymptotic expansions for integrals of the type (2.6) for the fairly wide class of $\phi(t)$ of the form $t^\lambda g(t)$, where $g(t)$ is a function which possesses a Taylor series expansion about $t = 0$, with $g(0) \neq 0$, and λ is real and such that $\lambda > -1$ (to ensure convergence of the integral at

† See, for example, Watson, G. N. (1952). *Theory of Bessel functions*. Cambridge University Press.

$t = 0$). If $g(t)$ has a zero of order r at $t = 0$, we simply incorporate the t^r in the t^λ and define a new $g(t)$ such that $g(0) \neq 0$. We thus look for an asymptotic expansion of

$$f(x) = \int_0^T e^{-xt} t^\lambda g(t) \, dt, \quad x \to \infty, \tag{2.7}$$

for x real and positive and T some positive number, which may be infinite, with $g(t)$ and $\lambda > -1$ as described above. So that the integral (2.7) converges if $T \to \infty$, we require $|g(t)| < Ke^{ct}$ in $0 \leq t \leq T$ for some positive constants K and c ($< x$, of course).

Since $g(t)$ has a Taylor series about $t = 0$ we have

$$\left. \begin{array}{l} g(t) = \sum_{n=0}^{\infty} \frac{g^{(n)}(0)}{n!} t^n = \sum_{n=0}^{\infty} a_n t^n = \sum_{n=0}^{N} a_n t^n + r_N(t), \\ |r_N(t)| < L\, t^{N+1}, \text{ for } |t| < R \end{array} \right\} \tag{2.8}$$

for some finite radius of convergence R and finite constant L, and where the a_n are defined by (2.8). At this stage consider $T < R$ for simplicity: this restriction is *not* necessary for the result. Using (2.8), (2.7) becomes

$$f(x) = \int_0^T e^{-xt} t^\lambda \sum_{n=0}^{N} a_n t^n \, dt + \int_0^T e^{-xt} t^\lambda r_N(t) \, dt. \tag{2.9}$$

But

$$\left| \int_0^T e^{-xt} t^\lambda r_N(t) \, dt \right| < L \int_0^T e^{-xt} t^{\lambda+N+1} \, dt \tag{2.10}$$

and, on changing variables from t to $\tau = xt$ in the right-hand integral, we have

$$\int_0^T e^{-xt} t^{\lambda+N+1} \, dt = x^{-(\lambda+N+2)} \int_0^{xT} e^{-\tau} \tau^{\lambda+N+1} \, d\tau$$

$$= x^{-(\lambda+N+2)} \int_0^{\infty} e^{-\tau} \tau^{\lambda+N+1} \, d\tau -$$

$$-x^{-(\lambda+N+2)} \int_{xT}^{\infty} e^{-\tau} \tau^{\lambda+N+1} \, d\tau$$

$$= x^{-(\lambda+N+2)} \Gamma(\lambda+N+2) -$$

$$-x^{-(\lambda+N+2)} \int_{xT}^{\infty} e^{-\tau} \tau^{\lambda+N+1} \, d\tau.$$

On setting $\tau = xT(1+u)$ in the last integral, and using the fact that $(1+u)^a < e^{au}$ for any $a > 0$ and $u > 0$,

$$x^{-(\lambda+N+2)} \int_{xT}^{\infty} e^{-\tau} \tau^{\lambda+N+1} \, d\tau = T^{\lambda+N+2} e^{-xT} \int_{0}^{\infty} e^{-xTu} (1+u)^{\lambda+N+1} \, du$$

$$< T^{\lambda+N+2} e^{-xT} \int_{0}^{\infty} e^{-xTu} e^{(\lambda+N+1)u} \, du$$

$$= T^{\lambda+N+2} e^{-xT} \frac{1}{xT-(\lambda+N+1)}$$

$$= T^{\lambda+N+2} e^{-xT} \frac{1}{xT}\left(1+\frac{\lambda+N+1}{xT} + \cdots\right)$$

$$\sim T^{\lambda+N+1} \frac{e^{-xT}}{x}.$$

Thus, using this result, we have

$$\int_{0}^{T} e^{-xt} t^{\lambda+N+1} \, dt = x^{-(\lambda+N+2)} \Gamma(\lambda+N+2) + o(e^{-xT})$$
$$= x^{-(\lambda+N+2)} \Gamma(\lambda+N+2) + o(x^{-(\lambda+N+2)})$$

$$\text{as } x \to \infty. \tag{2.11}$$

Now, using (2.11) in (2.10),

$$\int_{0}^{T} e^{-xt} t^{\lambda} r_N(t) \, dt = O(x^{-(\lambda+N+2)}) \text{ as } x \to \infty. \tag{2.12}$$

Further, using the estimate in (2.11), the first integral in (2.9) becomes

$$\int_{0}^{T} e^{-xt} t^{\lambda} \sum_{n=0}^{N} a_n t^n \, dt = \sum_{n=0}^{N} a_n \left\{\left(\int_{0}^{\infty} - \int_{T}^{\infty}\right) e^{-xt} t^{\lambda+n} \, dt\right\}$$

$$= \sum_{n=0}^{N} a_n \Gamma(\lambda+n+1) x^{-(\lambda+n+1)} +$$

$$+ o(x^{-(\lambda+N+1)}) \text{ as } x \to \infty,$$

which, on substituting, together with (2.12), into (2.9), finally gives

$$f(x) = \sum_{n=0}^{N} a_n \Gamma(\lambda+n+1) x^{-(\lambda+n+1)} + O(x^{-(\lambda+N+2)}),$$

and so with this $f(x)$ in (2.7) for all N,

$$\int_0^T e^{-xt} t^\lambda g(t) \, dt \sim \sum_{n=0}^\infty \frac{g^{(n)}(0) \, \Gamma(\lambda+n+1)}{n! \, x^{\lambda+n+1}} \quad \text{as } x \to \infty,$$

$$(2.13)$$

which is *Watson's lemma*.

The result (2.13) also holds for complex x, z say, as long as $|\arg z| < \pi/2$. The proof in this case is given (see also exercise 5) by Copson (1965): the only difference to the above is that a more careful estimate of (2.12) for $|\arg z| < \pi/2$ is required.

One interesting and important point to note about (2.13) is that *all* contributions to the asymptotic approximation as $x \to \infty$ come from the region near $t = 0$ irrespective of the order of the zero of $t^\lambda g(t)$, or, in other words, irrespective of how large λ may be: the upper limit T does not appear at all in the approximation (2.13). The T in (2.13) may in fact be replaced by infinity. This does not mean that $g(t)$ must have a Taylor series with infinite radius of convergence R. The procedure in the case where $T > R$ is to consider the integral from 0 to T as the sum of two integrals from 0 to $T_1 < R$ and from T_1 to T. To show that this second integral is negligible asymptotically, the bound on $g(t)$, namely, $|g(t)| < K e^{ct}$, is used (see exercise 4 or Copson (1965)).

As an example, let us consider, for real x,

$$f(x) = \int_0^\infty e^{-xt} \log(1+t^2) \, dt \text{ as } x \to \infty.$$

For $t^2 < 1$,

$$\log(1+t^2) = t^2(1 - \tfrac{1}{2}t^2 + \ldots),$$

and since $\log(1+t^2)$ satisfies the requirements on $t^\lambda g(t)$ we have, from (2.13) with $T = \infty$ (although its value is not relevant),

$$\int_0^\infty e^{-xt} \log(1+t^2) \, dt \sim \frac{2!}{x^3} - \tfrac{1}{2}\frac{4!}{x^5} + \ldots \quad .$$

Another example, which is more in the nature of an important extension of Watson's lemma, covers integrals of the type

$$f(x) = \int_{-\alpha}^\beta e^{-xt^2} \phi(t) \, dt,$$

$$(2.14)$$

where α and β are positive constants and $\phi(t)$ is restricted to be a function of the form $t^\lambda g(t)$, which ensures convergence of the integral for all α and β in a similar way to $t^\lambda g(t)$ in (2.7). To reduce (2.14) to the form (2.13) write

$$t = \tau^{\frac{1}{2}} \text{ in } 0 \leq t \leq \beta,$$
$$t = -\tau^{\frac{1}{2}} \text{ in } -\alpha \leq t \leq 0,$$
\qquad (2.15)

in which case (2.14) becomes

$$f(x) = \tfrac{1}{2} \int_0^{\beta^2} e^{-x\tau} \phi(\tau^{\frac{1}{2}}) \tau^{-\frac{1}{2}} d\tau + \tfrac{1}{2} \int_0^{\alpha^2} e^{-x\tau} \phi(-\tau^{\frac{1}{2}}) \tau^{-\frac{1}{2}} d\tau.$$
\qquad (2.16)

Suppose, for example, that

$$\phi(t) = \sum_{n=0}^{\infty} a_n t^n \text{ for } |t| < R,$$
\qquad (2.17)

then, applying Watson's lemma to each of the integrals in (2.16) using (2.13) (assuming, conveniently for the moment only, that $R > \max(\alpha, \beta)$),

$$f(x) = \tfrac{1}{2} \int_0^{\beta^2} e^{-x\tau} \tau^{-\frac{1}{2}}(a_0 + a_1 \tau^{\frac{1}{2}} + a_2 \tau + a_3 \tau^{\frac{3}{2}} + \ldots) d\tau +$$

$$+ \tfrac{1}{2} \int_0^{\alpha^2} e^{-x\tau} \tau^{-\frac{1}{2}}(a_0 - a_1 \tau^{\frac{1}{2}} + a_2 \tau - a_3 \tau^{\frac{3}{2}} + \ldots) d\tau$$

$$\sim \int_0^T e^{-x\tau} \tau^{-\frac{1}{2}}(a_0 + a_2 \tau + a_4 \tau^2 + \ldots) d\tau,$$
\qquad (2.18)

where T is any positive number and the odd powers in the two integrals can be cancelled asymptotically because of the unimportance of the upper limits β^2 and α^2 in the asymptotic approximations. By a similar procedure to that used to get the estimate (2.11) we have

$$\int_0^T e^{-x\tau} \tau^{-\frac{1}{2}} d\tau = \int_0^{\infty} e^{-x\tau} \tau^{-\frac{1}{2}} d\tau - \int_T^{\infty} e^{-x\tau} \tau^{-\frac{1}{2}} d\tau$$

$$= \frac{1}{x^{\frac{1}{2}}} \int_0^{\infty} e^{-s} s^{-\frac{1}{2}} ds + O\left(\frac{e^{-xT}}{x}\right) \text{ as } x \to \infty$$

$$= \frac{\Gamma(\frac{1}{2})}{x^{\frac{1}{2}}} + O\left(\frac{e^{-xT}}{x}\right) \text{ as } x \to \infty,$$

and similarly

$$\int_0^T e^{-xt} \tau^{(2n-1)/2} \, d\tau = \frac{\Gamma\left(\dfrac{2n+1}{2}\right)}{x^{(2n+1)/2}} + O\left(\frac{e^{-xT}}{x}\right) \text{ as } x \to \infty,$$

and so from (2.14) and (2.17) defining the a_n, (2.18) gives

$$f(x) = \int_{-\alpha}^{\beta} e^{-xt^2} \, \phi(t) \, dt \sim \sum_{n=0}^{\infty} a_{2n} \Gamma\left(\frac{2n+1}{2}\right) x^{-(2n+1)/2} \text{ as } x \to \infty$$

(2.19)

or, since

$$\Gamma(\tfrac{1}{2}) = \sqrt{\pi}, \ \Gamma\left(\frac{2n+1}{2}\right) = \frac{2n-1}{2} \Gamma\left(\frac{2n-1}{2}\right),$$

and so on, (2.19) becomes

$$\int_{-\alpha}^{\beta} e^{-xt^2} \, \phi(t) \, dt \sim \frac{\sqrt{\pi}}{x^{\frac{1}{2}}}\left(a_0 + \frac{a_2}{2x} + \frac{1 \cdot 3 \, a_4}{2^2 \, x^2} + \ldots\right) \text{ as } x \to \infty.$$

(2.20)

Again all of the contributions to the asymptotic approximation come from the neighbourhood of $t = 0$.

The fact that the contributions to the asymptotic expansions of integrals such as (2.7) and (2.14) all come from the region near $t = 0$, which is the value giving the maximum of $-xt$ and $-xt^2$ and hence the maximum of e^{-xt} and e^{-xt^2}, is indicative of a concept which is exploited in Laplace's method discussed below in §2.2. That the region near $t = 0$ becomes increasingly more important the larger x becomes, can be illustrated simply by looking, for example, at e^{-xt}

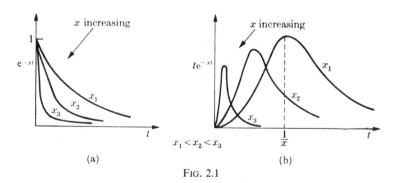

(a)   (b)

FIG. 2.1

and te^{-xt} as x increases: these functions are sketched respectively in Figs. 2.1(a) and 2.1(b).

In both cases, as x increases, the region around $t = 0$ which is important becomes smaller. If we had $e^{xh(t)}$ with $h'(t) < 0$ for all t, so that $t = 0$ gives a maximum $h(0)$ for $h(t)$, a similar kind of situation would arise, namely that the asymptotic contributions all come from the neighbourhood of $t = 0$. This situation and the important generalization called *Laplace's method* is discussed in §2.2.

Exercises

1. Using integration by parts, or term-by-term integration of an expanded integrand, prove that

   (i) $\int_0^\infty e^{-t}(z+t)^{-1} \, dt \sim \sum_0^\infty (-1)^n \, n! \, z^{-(n+1)}$ as $z \to \infty$ in the domain
   $|\arg z| < \pi$;

   (ii) $\int_0^\infty e^{-t}(1+zt)^{-1} \, dt \sim \sum_{n=0}^\infty (-1)^n \, n! \, z^n$ as $z \to 0$ in the domain
   $|\arg z| < \dfrac{\pi}{2}$;

   (iii) $\int_x^\infty e^{it} \, t^{-m} \, dt \sim \dfrac{ie^{ix}}{x^m} \sum_{n=0}^\infty \dfrac{(-1)^n \, i^n \, (m+n-1)!}{(m-1)! \, x^n}$
   for x real and large and m a positive integer;

   (iv) $e^{-x} \int_0^\infty \dfrac{e^{-t}}{x+t} \, dt = \dfrac{e^{-x}}{x} \int_0^\infty \dfrac{e^{-t}}{1+t/x} \, dt$

   $$= \dfrac{e^{-x}}{x} \int_0^\infty e^{-t} \left\{ 1 - \dfrac{t}{x} + \ldots + \dfrac{(-1)^{n-1} \, t^{n-1}}{x^{n-1}} + \right.$$
   $$\left. + \dfrac{(-1)^n \, t^n}{x^n \, (1+t/x)} \right\} \, dt$$

   $$\sim e^{-x} \left(\dfrac{1}{x} - \dfrac{1}{x^2} + \dfrac{2!}{x^3} - \ldots \right) \text{ as } x \to \infty,$$

   which is the asymptotic power series for Ei(x).

2. Defining the error function erf x by

   $$\operatorname{erf} x = \dfrac{2}{\sqrt{\pi}} \int_0^x e^{-t^2} \, dt, \; x \text{ real},$$

   obtain one asymptotic expansion as $x \to 0$ and one as $x \to \infty$.

3. Use Watson's lemma to obtain the first few terms in the asymptotic expanions for large x, of

(i) $\int\limits_0^\pi e^{-xt} t^{-1} \sin t \, dt$;   (ii) $\int\limits_0^{\pi/4} e^{-xt} (1 + \cos t)^{\frac{1}{2}} \, dt$.

4. Prove that Watson's lemma, equation (2.13), holds when the upper limit $T > R$, the radius of convergence of the Taylor series for $g(t)$, by using the bound $|g(t)| < Ke^{ct}$, for constants K and c, which holds for all $0 \leqq t \leqq T$.

5. Prove that Watson's lemma (2.13) holds for complex x, z say, in the domain $|\arg z| < \pi/2$, by carefully discussing the bound on the remainder integral (2.10) with x replaced by z.

6. Obtain the first few terms in the asymptotic expansion of $\int\limits_0^\infty e^{-xt^2} \sin t \, dt$ as $x \to \infty$.

7. Find an asymptotic approximation for $\int\limits_0^1 e^{-xt^3} \, dt$ as $x \to \infty$.

2.2.   Laplace's method

We now consider real integrals of the form $\int\limits_\alpha^\beta \phi(x, t) \, dt$, where the range of integration may be finite or infinite, $\phi(x, t)$ is a function of a real parameter x and we wish to find an asymptotic approximation as x tends to some limiting value, which to be specific we shall take to be infinity. The major contributions to the integral come from the neighbourhoods of the points of maximum $\phi(x, t)$ in the range of integration. If the largest contribution becomes progressively more dominant the larger x becomes, then we would expect the asymptotic approximation to the integral as $x \to \infty$ to come from the integration of $\phi(x, t)$ in the neighbourhood of this maximum value. In this situation we exploit the fact that $\phi(x, t)$ can be expressed, in such a neighbourhood, in terms of simpler functions, which can be easily integrated or evaluated asymptotically. This is essentially the idea behind Laplace's method for obtaining asymptotic approximations to such integrals. We have already seen in §2.1 how this occurs in two special cases, illustrated in Fig. 2.1, where the major contribution to the integral, in each case, comes from the integration over a neighbourhood of $t = 0$ and the contribution from this region becomes more dominant the larger x becomes. A typical $\phi(x, t)$ which is amenable to Laplace's technique in general is illustrated in Fig. 2.2.

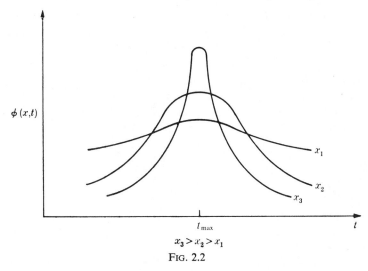

$\phi(x,t)$

x_1

x_2

x_3

t_{max}

t

$x_3 > x_2 > x_1$

FIG. 2.2

The most convenient form of the integrand $\phi(x, t)$ to discuss Laplace's method is that used by Laplace† who considered integrals of the type

$$f(x) = \int_{\alpha}^{\beta} g(t)\, e^{xh(t)}\, dt, \qquad (2.21)$$

where x is real and positive, $g(t)$ is a real continuous function, and $h(t)$, $h'(t)$ and $h''(t)$ are real and continuous in $\alpha \leqq t \leqq \beta$, where α and β are real. We look for an asymptotic approximation as $x \to \infty$. As related to (2.21) the essence of Laplace's idea is that the major contribution to the integral and hence to its asymptotic approximation as $x \to \infty$ will come from the neighbourhood of the point in $\alpha \leqq t \leqq \beta$, where $h(t)$ has its maximum value. If there are several maxima then the asymptotic approximation may have contributions from each of them: it will depend on the relative values of $h(t)$ at the maxima. If $h'(t) < 0$, for example, in $\alpha \leqq t \leqq \beta$, then $h(\alpha)$ is the maximum of $h(t)$. The examples discussed in §2.1 above under Watson's lemma are specific examples of (2.21) with $h(t) = -t$ in (2.7) and $h(t) = -t^2$ in (2.14): in the latter case $t = 0$ is a genuine maximum in that $h'(0) = 0$, $h''(0) < 0$. Watson's lemma *proved* that the asymptotic expansion as $x \to \infty$ came from evaluating the

---

† de Laplace, P. S. (1820). *Théorie analytique des probabilités*. Paris.

integral in the neighbourhood of $t = 0$ where the $h(t)$ had its maximum value. By a suitable change of variable, examples of which are given below, (2.21) can be written in the form of (2.7) or (2.14), and we then have Watson's lemma both to give and justify the asymptotic expansion. We shall not, therefore, prove Laplace's method in its own right: a proof not specifically using Watson's lemma is given in Copson's (1965) book. In this section we shall discuss the formal procedure of Laplace's method and obtain the asymptotic approximation for a fairly general class of integrals, taking note of particular situations which frequently arise in practice.

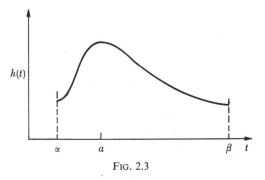

FIG. 2.3

When $h(t)$ has a maximum at an internal point $t = a$, say, where $\alpha < a < \beta$, $h'(a) = 0$: a typical $h(t)$ is illustrated in Fig. 2.3. This situation can be reduced immediately to the class in which the maximum is at an end-point of the range of integration by changing the variable so that $t = a$ is the origin, and then considering the two integrals which represent the integration from α to a and from a to β separately. As we shall see below ((2.32) *et seq.*) it is not necessary to split the integral (2.21) like this, but we do it here so that in what follows we may use Watson's lemma immediately. We write

$$\int_\alpha^\beta g(t)\, e^{xh(t)}\, dt = \int_\alpha^a g(t)\, e^{xh(t)}\, dt + \int_a^\beta g(t)\, e^{xh(t)}\, dt$$

$$= \int_0^{a-\alpha} g_1(\tau)\, e^{xh_1(\tau)}\, d\tau + \int_0^{\beta-a} g_2(\tau)\, e^{xh_2(\tau)}\, d\tau,$$

$$(2.22)$$

where $g_1(\tau) = g(a-\tau)$, $h_1(\tau) = h(a-\tau)$, $g_2(\tau) = g(a+\tau)$ and $h_2(\tau) = h(a+\tau)$. Each integral in (2.22) now has the maximum of the exponential at $\tau = 0$. We thus consider, without loss of generality, integrals of the form

$$f(x) = \int_0^T g(t)\, e^{xh(t)}\, dt, \tag{2.23}$$

where $h(0)$ is the maximum $h(t)$ in the range $0 \le t \le T$, where T is a positive number. A typical $h(t)$ is illustrated in Fig. 2.4.

At this stage it should be mentioned again that if, in the general integral (2.21), $h(t)$ has several relative maxima as in Fig. 2.5, we simply split the integral up into several integrals, each of which has a

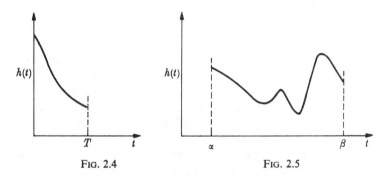

FIG. 2.4   FIG. 2.5

single maximum, and so again we need consider only one class (2.23). In the final analysis, the collection of asymptotic approximations from each integral have to be compared but, as will become clear below, the only really relevant cases are those in which $h(t)$ attains the *same* value at the different points of maxima. In these situations the final ordering of the asymptotic expansion will depend crucially on the $g(t)$ at the various points of maximum $h(t)$.

Returning to (2.23) and Fig. 2.4, the origin may be a genuine maximum with $h'(0) = 0$, $h''(0) < 0$ or it may not, in which case $h'(0) < 0$: examples of both have been discussed in §2.1. We shall consider separately the two situations (i) $h'(0) = 0$, $h''(0) < 0$ and (ii) $h'(0) < 0$.

Case (i) We consider (2.23) here with $h(0) > h(t)$ for all $0 < t \le T$ with $h'(0) = 0$, $h''(0) < 0$, and $g(t)$ and $h''(t)$ real continuous functions on $0 \le t \le T$. We wish to find an asymptotic approximation for $f(x)$ as $x \to \infty$. Throughout we should keep in mind that as $x \to \infty$ the dominant part of the integral—and hence its asymptotic approximation—comes from the immediate neighbourhood of $t = 0$. In some interval sufficiently close to $t = 0$, say when $0 \leq t \leq \delta < T$, $h''(t) < 0$ since $h''(t)$ is continuous and $h(0)$ is a genuine maximum. In this interval, using the mean value theorem in $0 \leq t \leq \delta$, $h(t) - h(0) = \frac{1}{2}t^2 h''(\xi)$, where $h''(\xi) < 0$ and $0 < \xi < \delta$. This suggests that in seeking an asymptotic approximation as $x \to \infty$ an appropriate new variable to use, s say, is defined by

$$h(t) - h(0) = -s^2. \tag{2.24}$$

With this the exponential in (2.23) becomes

$$e^{xh(t)} = e^{xh(0)} e^{-xs^2}, \tag{2.25}$$

and so the integral (2.23) in terms of s then belongs to the class (2.20) already discussed in §2.1 with s for t, and for which we have a definite proof, using Watson's lemma, that an asymptotic power series can be obtained as $x \to \infty$. From the procedural point of view, the next step is to replace $g(t)$ in (2.23) by its form as a function of s using (2.24). Since the dominant part of the integral as $x \to \infty$ comes from the neighbourhood of $t = 0$, the first approximation to $g(t)$ as a function of s is $g(0)$. So as to cover this approximation and to indicate generally how to obtain higher-order terms in the expansion, we consider $g(t)$ to be not only continuous but also to possess a Taylor series in the neighbourhood of $t = 0$. Thus we write

$$g(t) = g(0) + tg'(0) + \tfrac{1}{2}t^2 g''(0) + \ldots \tag{2.26}$$

valid for some finite radius of convergence. Further, if we are in a small neighbourhood of $t = 0$, expanding the left-hand side of (2.24), a first approximation for t in terms of s is obtained from

$$\tfrac{1}{2}t^2 h''(0) + \ldots = -s^2$$

as

$$t = \left\{ \frac{-2}{h''(0)} \right\}^{\frac{1}{2}} s + O(s^2). \tag{2.27}$$

Thus

$$g(t) = g(0) + g'(0) \left\{ \frac{-2}{h''(0)} \right\}^{\frac{1}{2}} s + O(s^2). \tag{2.28}$$

We now change the variable from t to s in (2.23), using (2.24) with (2.27) and (2.28) and, since the asymptotic approximation as $x \to \infty$ is given by the integration in the neighbourhood of $t = 0 = s$, we have

$$f(x) \sim e^{xh(0)} \left\{\frac{-2}{h''(0)}\right\}^{\frac{1}{2}} \int_0^A e^{-xs^2} \{g(0) + O(s)\} \, ds, \qquad (2.29)$$

where A is simply some positive number which is related to T in (2.23), but which may be taken to be infinite since the asymptotic evaluation of the integral in (2.29) does *not* depend on it, as was proved in §2.1. We thus have from (2.29)

$$f(x) \sim e^{xh(0)} g(0) \left\{\frac{-2}{h''(0)}\right\}^{\frac{1}{2}} \int_0^\infty e^{-xs^2} \, ds + e^{xh(0)} O(\int_0^\infty se^{-xs^2} \, ds) + \ldots,$$

$$(2.30)$$

and so, finally, since

$$\int_0^A e^{-xs^2} \, ds \sim \frac{1}{2}\left(\frac{\pi}{x}\right)^{\frac{1}{2}} \quad \text{as } x \to \infty$$

plus exponentially small terms,

$$\int_0^T g(t) e^{xh(t)} \, dt = g(0) \left\{\frac{-\pi}{2xh''(0)}\right\}^{\frac{1}{2}} e^{xh(0)} + e^{xh(0)} \, O\left(\frac{1}{x}\right).$$

$$(2.31)$$

Evaluation of the second term on the right of (2.31) simply requires the expression for the $O(s^2)$ term in (2.27), because

$$dt = \left\{\left(\frac{-2}{h''(0)}\right)^{\frac{1}{2}} + O(s)\right\} ds,$$

so that with (2.28) the $O(s)$ term in (2.29) can be given exactly. Some of the exercises (parts of exercise 2) require this.

If we now consider in place of (2.23),

$$f(x) = \int_{-U}^T g(t) e^{xh(t)} \, dt \quad \text{as } x \to \infty, \qquad (2.32)$$

where T, U are positive numbers and $h(0)$ is the maximum of $h(t)$, with $h'(0) = 0$, $h''(0) < 0$, then exactly the same procedure as above results in the integrals in (2.30) being from $-\infty$ to ∞ in place of 0 to ∞, and so the *first* term in (2.31) is simply multiplied by 2. As mentioned above, it is clear in this case that it is not necessary to split the integral (2.21) into two, as was done in (2.22). The important point to note is that the second term is *not* $e^{xh(0)} O(1/x)$ as in (2.31) because $\int_{-\infty}^\infty se^{-xs^2} \, ds = 0$. Thus terms of $O(s^2)$ and $O(s^3)$ in (2.27), and $O(s^2)$ in (2.28) must be obtained, and in this case the first non-zero contribution to the second term is from $\int_{-\infty}^{\infty} s^2 e^{-xs^2} ds = \frac{1}{2}(\pi/x^3)^{\frac{1}{2}}$†. Thus the asymptotic approximation for (2.32) as $x \to \infty$ is

$$\int_{-U}^{T} g(t) e^{xh(t)} dt = g(0) \left\{\frac{-2\pi}{xh''(0)}\right\}^{\frac{1}{2}} e^{xh(0)} + e^{xh(0)} O(x^{-\frac{1}{2}}).$$

(2.33)

To return finally to the general integral (2.21), where $t = a$ is the point of maximum $h(t)$, with $\alpha < a < \beta$, we simply replace the zero by a in (2.33) to give, as $x \to \infty$,

$$\int_{\alpha}^{\beta} g(t) e^{xh(t)} dt = g(a) \left\{\frac{-2\pi}{xh''(a)}\right\}^{\frac{1}{2}} e^{xh(a)} + e^{xh(a)} O(x^{-\frac{1}{2}}).$$

(2.34)

Case (ii) We now consider (2.23) where $h(0)$ is the maximum $h(t)$ in $0 \leq t \leq T$, with $h'(0) < 0$. As $x \to \infty$ the dominant contribution to the integral still comes from the small, δ say, neighbourhood of $t = 0$; and here, by the mean value theorem, $h(t) - h(0) = h'(\eta)t$, where $h'(\eta) < 0$ and $0 < \eta < \delta < T$. Since $h'(0) < 0$ and, by continuity, $h'(t) < 0$ for a sufficiently small neighbourhood of $t = 0$, the appropriate change of variable is, in place of (2.24),

$$h(t) - h(0) = -s, \tag{2.35}$$

which gives, for t small,

$$t = \left\{\frac{-1}{h'(0)}\right\}s + O(s^2), \tag{2.36}$$

in place of (2.27). Proceeding in a similar way to case (i) above, the asymptotic approximation to $f(x)$ is then given, in place of (2.29), by

$$f(x) \sim \int_{0}^{A} \{g(0) + O(s)\} e^{xh(0)} e^{-xs} \left\{\frac{-1}{h'(0)}\right\} ds$$

$$\sim \left\{\frac{-g(0)}{h'(0)}\right\} e^{xh(0)} \int_{0}^{\infty} e^{-xs} ds + e^{xh(0)} O\left(\int_{0}^{\infty} se^{-xs} ds\right),$$

† The simplest way to get this is to differentiate $\int_{-\infty}^{\infty} e^{-xs^2} ds = \left(\frac{\pi}{x}\right)^{\frac{1}{2}}$ with respect to x, that is

$$\int_{-\infty}^{\infty} s^2 e^{-xs^2} ds = -\frac{d}{dx} \int_{-\infty}^{\infty} e^{-xs^2} ds = -\frac{d}{dx} \sqrt{\frac{\pi}{x}} = \frac{1}{2} \left(\frac{\pi}{x^3}\right)^{\frac{1}{2}}.$$

and so, in the situation in which $h'(0) \neq 0$,

$$\int_0^T g(t) \, e^{xh(t)} \, dt = \left\{ \frac{-g(0)}{xh'(0)} \right\} e^{xh(0)} + e^{xh(0)} \, O(x^{-2}) \text{ as } x \to \infty. \quad (2.37)$$

Here the second term requires the $O(s^2)$ term in (2.36) as well as the $O(s)$ term in (2.28). Again, these have to be found for some of the exercises (parts of exercise 2).

If we now return to the integral (2.21), where the end point $t = \alpha$ gives the maximum $h(t)$ in the interval $\alpha \leq t \leq \beta$ but $h'(\alpha) < 0$, we have in place of (2.37)

$$\int_\alpha^\beta g(t) \, e^{xh(t)} \, dt = \left\{ \frac{-g(\alpha)}{xh'(\alpha)} \right\} e^{xh(\alpha)} + e^{xh(\alpha)} \, O(x^{-2}) \text{ as } x \to \infty. \quad (2.38)$$

If $h(\beta)$ is the maximum then the above results are only trivially modified. In the case comparable to (ii), where now $h(\beta) > h(t)$ for all $\alpha \leq t < \beta$, $h'(\beta) > 0$, in place of (2.38) we have,

$$\int_\alpha^\beta g(t) \, e^{xh(t)} \, dt = \left\{ \frac{g(\beta)}{xh'(\beta)} \right\} e^{xh(\beta)} + e^{xh(\beta)} \, O(x^{-2}) \text{ as } x \to \infty. \quad (2.39)$$

There are many extensions and generalizations of Laplace's method but the basic idea is always the same, namely that the asymptotic approximation is obtained by examining the integral in the vicinity of the maximum of the integrand. An obvious necessary extension is that which covers the situation in which $g(t)$ has a zero at the point of maximum $h(t)$. In this case higher-order terms are required: the situation is comparable to that in Watson's lemma. Suppose the maximum of $h(t)$ is at $t = 0$ and $g(t) = t^n \phi(t)$, $\phi(0) \neq 0$, where n is a positive integer (if n is not an integer we would finally have to deal with a similar situation to that treated in §2.1). The first integral in (2.30) now involves $\int_0^\infty s^n \, e^{-xs^2} \, ds$ which is easily integrated by parts or, alternatively, by differentiating with respect to x the integral $\int_0^\infty e^{-xs^2} \, ds = \sqrt{(\pi/x)}$ if n is even, or $\int_0^\infty se^{-xs^2} \, ds = 1/(2x)$ if n is odd. Another generalization discussed in Copson's (1965) book deals with integrals involving $e^{h(x,t)}$ in place of $e^{xh(t)}$, in which case the position of the maximum can depend on x as well. Another situation of interest is where $h(t)$ is a maximum at $t = 0$, say, in $0 \leq t \leq T$ but $h'(0) = 0 = h''(0)$, $h'''(0) \neq 0$, for example (see exercise 4).

As a typical example of the type of integral covered in this section, let us find the asymptotic approximation of

$$f(x) = \int_0^{\pi/2} e^{x \cos t} \log (\lambda + \sin t) \, dt \text{ as } x \to \infty,$$

where λ is a real positive constant. Here $h(t) = \cos t$ and $h(0)$ is a genuine maximum, since $h'(0) = 0$, $h''(0) = -1 < 0$, and so it is an example of case (i) above and hence (2.31) gives the asymptotic approximation, which since $g(0) = \log \lambda$, is

$$\int_0^{\pi/2} e^{x \cos t} \log (\lambda + \sin t) \, dt \sim \left(\frac{\pi}{2x}\right)^{\frac{1}{2}} e^x \log \lambda \text{ as } x \to \infty$$

as long as $\lambda \ne 1$. If $\lambda = 1$ then $g(0) = 0$. In this case we have to go to higher-order terms, namely $O(s)$ in (2.29). In place of (2.28) we have

$$g(t) = g'(0) \left\{\frac{-2}{h''(0)}\right\}^{\frac{1}{2}} s + O(s^2)$$

$$= \frac{d}{dt} [\log (1 + \sin t)]_{t=0} \, 2^{\frac{1}{2}} s + O(s^2)$$

$$= 2^{\frac{1}{2}} s + O(s^2),$$

and so, in place of (2.29), we have

$$\int_0^{\pi/2} e^{x \cos t} \log (1 + \sin t) \, dt \sim e^x \sqrt{2} \int_0^A s e^{-xs^2} \{\sqrt{2} + O(s)\} \, ds,$$

$$\sim \frac{e^x}{x} \text{ as } x \to \infty.$$

The integrand in this example is a smooth function of *two* variables x and λ, and so we expect the asymptotic expansion of the integral to be a smooth function of both x and λ. The difference above between the $\lambda \ne 1$ and $\lambda = 1$ cases does *not* in any way represent a discontinuity in the integral, as a function of λ, in the sense that if $\lambda = 1 + \varepsilon$, for any $\varepsilon > 0$ however small, then the asymptotic expansion is $e^x (\pi/2x)^{\frac{1}{2}}$ whereas when $\varepsilon = 0$ it is e^x/x. To see this, we need to consider the expansion more carefully by evaluating it to at least two terms for $\lambda \ne 1$ and *then* consider the limit $\lambda \to 1$. Here the transformation from t to s for (2.27) is

$$t = \sqrt{2} \, s + \frac{\sqrt{2}}{12} s^3 + O(s^4),$$

and for (2.28) the expansion in s is

$$g(t) = \log \lambda + \frac{\sqrt{2}}{\lambda} s + O(s^2),$$

and so

$$\int_0^{\pi/2} e^{x \cos t} \log (\lambda + \sin t)\, dt \sim e^x \int_0^{\infty} e^{-xs^2} \left\{ \sqrt{2} \log \lambda + s \frac{2}{\lambda} + \ldots \right\} ds$$

$$= e^x \left\{ \left(\frac{\pi}{2x} \right)^{\frac{1}{2}} \log \lambda + \frac{1}{x} \frac{1}{\lambda} + O(x^{-\frac{3}{2}}) \right\},$$

which covers *both* of the above cases. The smooth transition from $\lambda \neq 1$ to $\lambda = 1$ for a given x is now seen on writing $\lambda = 1 + \varepsilon$ with $\varepsilon > 0$, expanding for $\varepsilon \ll 1$, and rewriting the last expansion as one in ε which gives

$$\int_0^{\pi/2} e^{x \cos t} \log (\lambda + \sin t)\, dt \sim e^x \left[\frac{1}{x} + \varepsilon \left\{ \left(\frac{\pi}{2x} \right)^{\frac{1}{2}} - \frac{1}{x} \right\} + O(\varepsilon^2) \right]$$

as $x \to \infty$, $\varepsilon \to 0$. The point is that as $\lambda \to 1, \log \lambda \to 0$ sufficiently to make the multiple of $O(x^{-\frac{1}{2}})$ term smaller than the $O(x^{-1})$ term: e^x is a factor in each case. This example illustrates an important point which is connected with the order of the limits in asymptotic analysis when there is more than one variable tending to a given limit. In this example, for a *given* ε however small, the expansion is dominated, for x sufficiently large, by $e^x (\pi/2x)^{\frac{1}{2}}$, whereas for a *given* x, however large but finite, there always exists an ε such that the expansion is dominated by e^x/x. This is a way of saying that $\lim_{\varepsilon \to 0} \lim_{x \to \infty}$ is *not* the same as $\lim_{x \to \infty} \lim_{\varepsilon \to 0}$ from an asymptotic expansion point of view.

As another example of Laplace's method, we consider the gamma function $\Gamma(x)$ defined for real x by (2.2) and look for an asymptotic approximation as $x \to \infty$. This problem was raised in §2.1 where it was pointed out that integration by parts was not a suitable method. It is convenient to consider

$$\Gamma(x+1) = \int_0^{\infty} e^{-t} t^x\, dt = \int_0^{\infty} e^{-t+x \log t}\, dt \text{ as } x \to \infty,$$

which may be put in a form suitable for Laplace's method, on setting

$t = x\tau$ in the last integral, to give

$$\Gamma(x+1) = x \int_0^\infty e^{-x\tau + x \log \tau + x \log x} \, d\tau$$

$$= x^{x+1} \int_0^\infty e^{x(-\tau + \log \tau)} \, d\tau.$$

This integral is in the form of (2.21) with $g(\tau) = 1$ and $h(\tau) = -\tau + \log \tau$, which has a maximum at an internal point $\tau = 1$, with $h(1) = -1$ and $h''(1) = -1 < 0$, and so (2.34) immediately gives

$$\Gamma(x+1) \sim x^{x+1} e^{-x} \left(\frac{2\pi}{x}\right)^{\frac{1}{2}},$$

and so

$$\Gamma(x+1) \sim x^x \sqrt{(2\pi x)} \, e^{-x} \text{ as } x \to \infty. \tag{2.40}$$

If x is an integer n, say, then $\Gamma(n+1) = n!$ and (2.40) thus shows that

$$n! \sim n^n \sqrt{(2\pi n)} \, e^{-n} \text{ as } n \to \infty,$$

which is known as Stirling's formula.

In this section we have considered only real integrals. Most of the results can be extended to certain complex integrals, for example, when x is complex. To be able to deal with Laplace, Fourier and general integral transforms, we must be able to deal with more general complex integrals than those, like the above, with simply x complex. This we do in the next chapter where we discuss what is called the *method of steepest descents* or the *saddle-point method*, which with its various generalizations is probably the most general and important technique available for obtaining asymptotic approximations of integrals.

Exercises

1. Use Laplace's method to verify the given asymptotic approximations as $x \to \infty$ of the following:

   (i) $\displaystyle\int_0^{\pi/4} e^{x \cos t} \cos nt \, dt \sim \left(\frac{\pi}{2x}\right)^{\frac{1}{2}} e^x$, n an integer;

   (ii) $\displaystyle\int_{-1}^1 e^{-x \cosh t} \, dt \sim \left(\frac{2\pi}{x}\right)^{\frac{1}{2}} e^{-x}$;

   (iii) $\displaystyle\int_0^1 e^{-xt^{\frac{1}{2}}} \frac{\cos t}{t^{\frac{1}{2}}} \, dt \sim \frac{2}{x}$;

(iv) $\int\limits_0^1 t^x \cos^n \pi t \, dt \sim \dfrac{(-1)^n}{x}$;

(v) $\int\limits_0^1 e^t \, t^x (1+t^2)^{-x} \, dt \sim \left(\dfrac{\pi}{2}\right)^{\frac{1}{2}} e \, x^{-\frac{1}{2}} 2^{-x}$.

2. Use Laplace's technique or, in some cases, Watson's lemma, to obtain the first two (except for (iii)) terms in the asymptotic expansion as $x \to \infty$ of the following:

(i) $\int\limits_0^\infty e^{-xt} \cos t \, dt$; (ii) $\int\limits_0^\infty e^{-xt} \sin t \, dt$;

(iii) $\int\limits_0^1 e^{x(t-t^2)} \, dt$; (iv) $\int\limits_0^\infty e^{-x \cosh^2 t} \, dt$;

(v) $\int\limits_{-\infty}^\infty e^{x(t-e^t)} \, dt$; (vi) $\int\limits_{-\infty}^\infty e^{-xt^2} \log(1+t^2) \, dt$.

3. Obtain the first asymptotic term of $\int\limits_0^T g(t) \, e^{xh(t)} \, dt$ as $x \to \infty$, where $T > 0, g(t)$ is continuous and $g(t) = t^n \phi(t)$, $\phi(0) \neq 0$, and n is a positive integer in the two cases:

(i) $h'(0) = 0$, $h''(0) < 0$, and $h(0)$ is the maximum $h(t)$ in $0 \leq t \leq T$;

(ii) $h(0)$ is the maximum $h(t)$ in $0 \leq t \leq T$ and $h'(0) < 0$.

In case (i) note the differences between n odd and n even.

4. Extend Laplace's method to obtain the asymptotic approximation as $x \to \infty$ of $\int\limits_0^\infty e^{-xt^3} g(t) \, dt$, where $g(t)$ is continuous and $g(0) \neq 0$.

5. Using the fact that $\int\limits_0^\infty e^{-nt} t^s \, dt = s! \, n^{-(s+1)}$ for s an integer, show that

$$\sum_{s=0}^n \binom{n}{s} s! \, n^{-s} \sim \left(\dfrac{\pi n}{2}\right)^{\frac{1}{2}} \quad \text{as } n \to \infty,$$

where $\dbinom{n}{s}$ are the binomial coefficients.

3

Method of steepest descents

3.1. Method of steepest descents

THE method of steepest descents or saddle-point method is essentially a generalization of Laplace's method to integrals in the complex plane. The method originated with Riemann[†]. It was developed in its present form independently by Debye[‡]. Various extensions and rigorous proofs of some of the procedures have been given since their work and some of these are in the books cited. The basic idea and method which still has the widest applicability, are discussed in this section, with several illustrative worked examples being given in §3.2.

Although the remarks below on the philosophy of the method apply to complex integrals in general, the method is generally only of immediate practical use for integrals which can be put in a form which belongs to the specific class of integrals we now consider, namely,

$$f(\lambda) = \int_C g(z)\, e^{\lambda h(z)}\, dz, \qquad (3.1)$$

where C is some contour in the complex plane, $g(z)$ and $h(z)$, which are independent of λ, are analytic functions of z in some domain of the complex plane which contains C, and λ is a real positive number. The problem is to find an asympotoic approximation for $f(\lambda)$ with λ large. Naturally we consider only those integrals in (3.1) which exist and are finite. The method is applicable to more general inte-

---

† Riemann, B. (1892). *Gesammelte mathematische werke* (2nd ed.). Reprinted Dover Publications Inc., New York, 1953.

‡ Debye, P. (1909). *Math. Ann.* **67**, 535–558. (See also, *Collected papers* (1954). Interscience, New York.)

grands, an important example being that in which $h(z)$ is a function of λ as well as z (compare with the similar extension to Laplace's method): the reader is referred to the fuller discussions of the method in the books cited.

Note that $g(z)$ and $h(z)$ are not necessarily analytic in the whole complex plane. They may, and in practice frequently do, have isolated singularities including branch points, the branch lines for which must be carefully noted when using the saddle-point method. In the illustrative examples given below in §3.2, cases in which $g(z)$ and/or $h(z)$ have singularities are included. Further, choosing λ to be real and positive involves no loss of generality since if a similar integral to (3.1) arises in which λ is complex and $\lambda \rightarrow \infty$ along a ray $\lambda = |\lambda| e^{i\alpha}$, we simply incorporate the $e^{i\alpha}$ into the $h(z)$ and we have again an integral like (3.1) with the effective λ real and positive. The case when λ is real and negative is covered, of course, by taking $\alpha = \pi$ in the complex λ situation.

To get some heuristic understanding of the basic idea behind the method of steepest descents and to see the connection with Laplace's method, we first consider a limit inequality argument for $|f(\lambda)|$ from (3.1). As a convenient preliminary, we introduce the functions ϕ and ψ by writing

$$h(z) = \phi + i\psi, \quad \phi = \text{Rl } h(z), \quad \psi = \text{Im } h(z). \tag{3.2}$$

If the contour C, which may be finite or infinite, joins the points $z = a$ and $z = b$, we have from (3.1), with (3.2),

$$|f(\lambda)| \leq \int_{s_a}^{s_b} |g(z) \, e^{\lambda h(z)}| \, ds$$

$$\leq \int_{s_a}^{s_b} |g| \, e^{\lambda \phi} \, ds, \tag{3.3}$$

where $ds = |dz|$ and s_b and s_a correspond to the end points $z = b$ and $z = a$ of the contour. If, for the purposes of this preliminary discussion only, $\int_C g(z) \, dz$ is absolutely convergent (that is, $\int_{s_a}^{s_b} |g| \, ds$ is convergent), then for λ large the integral in (3.3), using the results of §2.2, is $O(e^{\lambda \phi})$, except for the usual multiplicative algebraic terms like $\lambda^{-\frac{1}{2}}$, λ^{-1}, and so on. If the contour is of finite length L, say, then in place of (3.3) we have

$$|f(\lambda)| < L \max_L (|g(z)| \, e^{\lambda \phi}), \tag{3.4}$$

where $\max_L$ means the maximum of $|g| \, e^{\lambda \phi}$ on the path C of length L.

In (3.3) and (3.4) the most important contribution to the asymptotic approximation for $|f(\lambda)|$ as $\lambda \to \infty$ must come from the neighbourhood of the point of maximum ϕ: in (3.3), particularly, this is reminiscent of Laplace's method. We now exploit the fact that the contour C in (3.1) can be deformed, by Cauchy's theorem, into other contours, at least in the domain of analyticity of $h(z)$ and $g(z)$. If g has an isolated pole singularity, for example, we can still deform the contour into another which involves crossing such a singularity, if we use the theory of residues appropriately. Branch points and lines are a different matter as illustrated in the example below in §3.2.

We now deform the path C so that it not only passes through the point $z = z_0$, say, where $\phi = \text{Rl } h(z)$ has its maximum value, but also along it ϕ drops off on either side of its maximum as rapidly as possible. In this way the largest value of $e^{\lambda\phi}$ as $\lambda \to \infty$ will be concentrated in a small section of the contour. This specific path through the point of maximum $e^{\lambda\phi}$ will be a path of *steepest descents*. In the case of (3.4) the length L of the path also varies in this contour-deforming exercise. When a path is close to the optimal one in the steepest descents sense, a very small variation in the path, and hence its length L, can drastically change the variation of $e^{\lambda\phi}$ in the neighbourhood of its maximum when λ is large. Thus, if we were interested only in $|f(\lambda)|$ as $\lambda \to \infty$ we would choose a path which made ϕ and hence $e^{\lambda\phi}$ behave in the above manner, and then use Laplace's method. However, since we are interested in $f(\lambda)$ and not just its modulus as $\lambda \to \infty$, we must be more specific, keeping in mind the above discussion.

If we take any path through z_0, the point giving the maximum ϕ, the imaginary part ψ of $h(z)$ gives an oscillatory contribution $e^{i\lambda\psi}$ to the integrand. These oscillations are increasingly more rapid (or dense) the larger λ becomes. This jeopardizes the whole procedure unless a contour can be chosen which reduces the effect of the oscillations on the integrand *in the vicinity* of z_0, so that we can use the above idea. An appropriate path to choose to overcome this problem is one in which $\psi = \text{Im } h(z) = $ constant, in the vicinity of z_0, so that there are no oscillations from $e^{i\lambda\psi}$ near the point of maximum $e^{\lambda\phi}$. This is exactly what we do.

We return now to $f(\lambda)$ given by (3.1) with the given conditions on g, h, λ, and the contour C. For illustrative purposes, let us use Cartesian coordinates x, y and write

$$z = x+iy, \quad h(z) = \phi(x, y)+i\psi(x, y). \tag{3.5}$$

A relative maximum of ϕ is given by $z_0 = x_0 + iy_0$, a solution of $\nabla\phi = 0$, where ∇ is the usual gradient operator which in the situation here is $\mathbf{i}\,\partial/\partial x + \mathbf{j}\,\partial/\partial y$, where $\mathbf{i}$ and $\mathbf{j}$ are unit vectors in the x- and y-directions. Since ϕ and ψ are the real and imaginary parts of $h(z)$, an analytic function of z, they satisfy the Cauchy–Riemann equations which here are

$$\frac{\partial\phi}{\partial x} = \frac{\partial\psi}{\partial y}, \quad \frac{\partial\phi}{\partial y} = -\frac{\partial\psi}{\partial x}. \tag{3.6}$$

Thus z_0 is also a solution of $\nabla\psi = 0$ and hence

$$h'(z) = \phi_x + i\psi_x = \phi_x - i\phi_y = 0 \text{ when } z = z_0 = x_0 + iy_0. \tag{3.7}$$

But from (3.6) (or generally because $\phi + i\psi = h(z)$, which is analytic) ϕ and ψ are potential functions satisfying Laplace's equation

$$\Delta\phi = 0, \quad \Delta\psi = 0, \tag{3.8}$$

where Δ is the Laplacian operator which here is simply $\partial^2/\partial x^2 + \partial^2/\partial y^2$. However, by the maximum modulus theorem, ϕ and ψ *cannot* have a maximum (or a minimum) in the domain of analyticity of $h(z)$. The point z_0 is thus a *saddle-point* of ϕ and of ψ. Because of (3.7) we say that z_0 is a saddle-point of $h(z)$. Here we shall be concerned with saddle-points of order one, that is

$$h'(z_0) = 0, \quad h''(z_0) \neq 0. \tag{3.9}$$

The procedure to use for higher-order saddle-points (a saddle-point of order n has the first n derivatives zero at the point) is basically an extension of the method discussed below.

If we consider the surface given by $\phi = \phi(x, y)$ in the ϕ, x, y space, a typical saddle-point situation is illustrated in Fig. 3.1 where the point S' in the surface is the saddle-point corresponding to the point S at $z = z_0$ in the z-plane. The lower part of Fig. 3.1 illustrates the contour lines, that is the projections onto the z-plane of the intersections of the planes $\phi = $ constant and the three-dimensional surface $\phi = \phi(x, y)$ in the ϕ, x, y space. Primes on the points in the surface relief correspond to the unprimed points in the z-plane. For example, the constant ϕ plane which intersects the surface all along the curves $E'P'F'$ and $G'Q'H'$ in the upper half of Fig. 3.1 has, as its projection on the z-plane, the lines EPF and GQH respectively. From (3.7) the tangent plane at the saddle point S' on the surface is a constant ϕ-plane parallel to the z-plane: the intersection of it with the surface projects onto the lines ASD and BSC.

We now return to the question of the optimal path into which we must deform the contour C. To avoid the oscillation problem of $e^{i\lambda\psi}$, we choose a path along which $\psi = \text{Im } h(z) = $ constant. From the Cauchy–Riemann equations, since $\nabla\phi \cdot \nabla\psi = 0$, the lines $\phi = $ *constant* and $\psi = $ *constant* are *orthogonal* and so the lines along which ϕ changes most rapidly, that is the direction of $\nabla\phi$, are thus the lines $\psi = $ *constant*. If we choose the ψ-line which passes through the point z_0, where ϕ has its saddle-point, then this is in keeping with the necessary conditions that (i) along the optimal path ϕ has as rapid a variation as possible near its relative maximum and (ii) there are no oscillation contributions from $e^{i\lambda\psi}$. If we now look in detail at Fig. 3.1, we see, from the sketch of the contour lines (solid) of constant $\phi = \text{Rl } h(z)$ on the z-plane, that there are *two* lines $\psi = \text{Im } h(z) = $ *constant* which pass through the saddle-point S at z_0 and along which ϕ changes as rapidly as is possible. However, the dashed line through PSQ which lies in the shaded portion of the z-plane, corresponds *not* to the steepest *descent* path but to the steepest *ascent* path ($P'S'Q'$ on the surface) since the value of ϕ on it is such that $\phi(x, y) > \phi(x_0, y_0)$ except at z_0. What is more, the values of ϕ on such a line become unbounded far from the saddle point: this is easily seen in the three-dimensional relief part of Fig. 3.1 where, as mentioned above, the primed letters in the relief correspond to the same unprimed letters in the plane. Thus the original contour C, for which the integral for $f(\lambda)$ in (3.1) exists, *cannot* be deformed into this (the line PSQ) $\psi = \text{Im } h(z) = $ *constant* contour nor any other contour which lies in the shaded 'mountain' regions. If we now consider the dashed line $\psi = $ *constant* in Fig. 3.1 passing through MSN in the z-plane and relate it to the relief surface line $M'S'N'$, we see that for any $z = x+iy$ on this line $\phi(x, y) < \phi(x_0, y_0)$, except at $z = z_0$. Thus this $\psi = $ *constant* line through MSN is the line of *steepest descents* and the one into which C is to be deformed. Any line which lies in the unshaded 'valley' regions and passes through z_0, the point S, is a possible deformable contour to the optimal one. The use of the words 'mountain' and 'valley' regions are often used to denote the shaded and unshaded domains in the z-plane: keeping in mind the relief surface from which they came it is a convenient and obvious description.

It might be argued that a path which starts and finishes in the

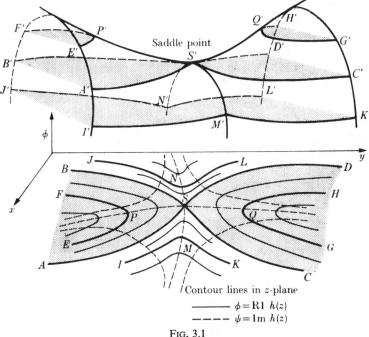

FIG. 3.1

valley regions and passes through not S, but Q, say, which lies in the shaded mountain region and hence higher up the saddle at Q in the relief surface, is better than the one we have chosen, since over part of such a path $\phi(x, y) > \phi(x_0, y_0)$: the integral still exists for such a path. However, in this situation we *cannot* stay on a single constant ψ-line and so we have an oscillation contribution from $e^{i\lambda\psi}$ to contend with, and the argument that the major contribution to the integral as $\lambda \to \infty$ comes from the region of maximum ϕ is no longer valid.

In practice the correct path of the two $\psi = constant$ lines through the saddle-point is obtained simply by considering $\mathrm{Rl}\ h(z) = \phi$ along both and choosing that in which $\phi(x, y) < \phi(x_0, y_0)$ except at z_0.

Up to now we have assumed that the original contour has one end in each of the two valleys so that it can be deformed into the optimal path of steepest descents. If it turns out that the contour lies wholly on one side of the saddle-point with both ends in the same valley, the method can still be used since the contour can still be deformed into lines of constant ψ and the steepest descents philosophy still applies. In §3.2 the worked example (iv) illustrates specifically how the method proceeds in such a case.

With the general procedure in mind, we now examine the problem analytically to obtain an asymptotic approximation for $f(\lambda)$ in (3.1) as $\lambda \to \infty$. It will be helpful in the following to keep Fig. 3.1 in mind. Near the saddle-point z_0, defined by (3.9) as the point where $h'(z_0) = 0$, we can expand $h(z)$ in a Taylor series, and so

$$h(z) = h(z_0) + \tfrac{1}{2}(z - z_0)^2 \, h''(z_0) + O((z - z_0)^3). \qquad (3.10)$$

If we write

$$h''(z_0) = a e^{i\alpha}, \quad a > 0, \quad z - z_0 = r e^{i\theta}, \quad r > 0 \qquad (3.11)$$

then (3.10), with ϕ and ψ from (3.5), becomes

$$\phi(x, y) + i\psi(x, y) = \phi_0 + i\psi_0 + \tfrac{1}{2}ar^2 \, e^{i(2\theta + \alpha)} + O(r^3), \qquad (3.12)$$

where we have written

$$\phi_0 = \phi(x_0, y_0), \quad \psi_0 = \psi(x_0, y_0), \qquad (3.13)$$

the values of ϕ and ψ at the saddle-point. Equating the real parts of (3.12), we have

$$\phi = \phi_0 + \tfrac{1}{2}ar^2 \cos(2\theta + \alpha) + O(r^3). \qquad (3.14)$$

From (3.14) there are *two* contour lines $\phi = \phi_0$ which, close enough to z_0, that is when r is small, are tangent to the *two orthogonal* lines which are solutions of $\tfrac{1}{2}ar^2 \cos(2\theta + \alpha) = 0$ for $a \neq 0$ and $r \neq 0$. These are the lines

$$\theta = \tfrac{1}{2}\left(\frac{\pi}{2} - \alpha\right) \quad \text{and its continuation} \quad \theta = \pi + \tfrac{1}{2}\left(\frac{\pi}{2} - \alpha\right),$$

and

$$\theta = \tfrac{1}{2}\left(-\frac{\pi}{2} - \alpha\right) \quad \text{and its continuation} \quad \theta = \pi + \tfrac{1}{2}\left(-\frac{\pi}{2} - \alpha\right). \qquad (3.15)$$

The constant $\phi = \phi_0$ lines are sketched in Fig. 3.2: they are essentially the lines given by (3.15) for z close enough to z_0. Figure 3.2 is an enlargement of the vicinity of S in the z-plane in Fig. 3.1. Thus, there are two ranges of θ around z_0, namely

$$\frac{\pi}{4} - \frac{\alpha}{2} < \theta < \frac{3\pi}{4} - \frac{\alpha}{2} \quad \text{and} \quad \frac{5\pi}{4} - \frac{\alpha}{2} < \theta < \frac{7\pi}{4} - \frac{\alpha}{2},$$

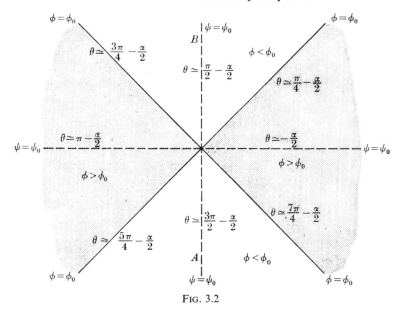

FIG. 3.2

in which $\phi < \phi_0$, as is seen on substituting such θ-ranges into (3.14): these are the unshaded valley regions in Fig. 3.2. Similarly shaded mountain regions are the domains in which $\phi > \phi_0$. If we now equate the imaginary parts in (3.12),

$$\psi = \psi_0 + \tfrac{1}{2}ar^2 \sin(2\theta + \alpha) + O(r^3). \tag{3.16}$$

The *two* curves $\psi = \psi_0$, along which ϕ changes most rapidly, are, from (3.16), tangent at $z = z_0$ to the orthogonal straight lines which are solutions of $\sin(2\theta + \alpha) = 0$ when a and $ar\ (\ll 1)$ are not zero, namely

$$\left.\begin{aligned} \theta &= -\frac{\alpha}{2} \quad \text{and its continuation } \theta = \pi - \frac{\alpha}{2}, \\[2mm] \theta &= \frac{\pi}{2} - \frac{\alpha}{2} \quad \text{and its continuation } \theta = \frac{3\pi}{2} - \frac{\alpha}{2}. \end{aligned}\right\} \tag{3.17}$$

The two $\psi = \psi_0$ lines illustrated in Fig. 3.2 by the dashed lines thus bisect the $\phi = \phi_0$ lines. Referring to Fig. 3.1 these $\psi = \psi_0$ lines are the two dashed lines through MSN (second of (3.17)) in the valleys and PSQ (first of (3.17)) in the mountain regions. The contour C can only be deformed into a path through z_0, which lies in the valleys, where $\phi < \phi_0$.

We now deform the contour C, assuming that it is a case where the end points lie one in each valley, so that it lies along the *steepest descents* path obtained from (3.16) on setting $\psi = \psi_0$; in which case, *on* it and near z_0,

$$\phi - \phi_0 = h(z) - h(z_0) = \tfrac{1}{2}(z - z_0)^2 \, h''(z_0) < 0, \qquad (3.18)$$

which is *real*: the imaginary parts of $h(z)$ and $h(z_0)$ cancel because $\psi = \psi_0$ on this path. Remembering now the Laplace procedure in §2.2, specifically equation (2.24), we introduce a new *real* variable τ by

$$h(z) - h(z_0) = -\tau^2, \quad \tau \text{ real}, \qquad (3.19)$$

which determines z as a function of τ, $z(\tau)$ say. The integral (3.1) for $f(\lambda)$ now becomes

$$f(\lambda) = e^{\lambda h(z_0)} \int_{-\tau_a}^{\tau_b} e^{-\lambda \tau^2} \, g(z(\tau)) \frac{dz}{d\tau} \, d\tau, \qquad (3.20)$$

where $\tau_a > 0$, $\tau_b > 0$ correspond to the end points of the original contour under the transformation (3.19). *On* the path of steepest descents τ is *real* and so the asymptotic expansion of (3.20) can now be obtained by Watson's lemma in §2.1 or Laplace's method in §2.2. Thus (3.20) gives

$$f(\lambda) \sim e^{\lambda h(z_0)} \int_{-\infty}^{\infty} e^{-\lambda \tau^2} \, g(z(\tau)) \frac{dz}{d\tau} \, d\tau \text{ as } \lambda \to \infty, \qquad (3.21)$$

where $z(\tau)$ is obtained from (3.19) when z lies on $\psi = \psi_0$, the steepest descent path. The fact that τ_a and τ_b are replaced by infinity in (3.21) is in keeping with the theory in §§2.1, 2.2 in which we proved that the asymptotic expansion as $\lambda \to \infty$ comes from the integration over a small region in the vicinity of the maximum of the exponential, which here is at $\tau = 0$. The only exercise now to be completed is the inversion of (3.19) on $\psi = \psi_0$ to give $z(\tau)$ and hence $g(z(\tau))$ and $dz/d\tau$ for (3.21). This we do by a systematic iterative procedure in the vicinity of z_0 in a similar way to that used in §2.2. The first term is easy to get. It is usually tedious to obtain the full series inversion which is generally of little practical interest. In the following and in the examples in §3.2 we shall obtain only the leading (or dominant) term in the asymptotic expansion.

From (3.19), expanding the left-hand side in a Taylor series about z_0 gives, since $h'(z_0) = 0$,

$$\tfrac{1}{2}(z - z_0)^2\, h''(z_0) + O((z - z_0)^3) = -\tau^2, \tag{3.22}$$

and so

$$z - z_0 = \left\{ \frac{-2}{h''(z_0)} \right\}^{\frac{1}{2}} \tau + O(\tau^2). \tag{3.23}$$

With $h''(z_0)$ complex, we must choose the appropriate branch of $\{-1/h''(z_0)\}^{\frac{1}{2}}$ when z lies on the steepest descent path. This is determined by the manner, in the sense of direction, in which we go through the branch point. Suppose, for example, that when C is deformed to lie along the steepest descent $\psi = \psi_0$ path in Fig. 3.2, the direction of integration is from A to B, that is, for z near z_0 from the arg $(z - z_0) = (3\pi/2) - \tfrac{1}{2}\alpha$ region to the arg $(z - z_0) = \tfrac{1}{2}\pi - \tfrac{1}{2}\alpha$ region. Then the appropriate branch to choose in (3.23) must be such that arg $\{-1/h''(z_0)\}^{\frac{1}{2}}$ gives $\tau > 0$ when z is in the upper B-valley region, that is when z is near z_0, arg $(z - z_0) = \tfrac{1}{2}\pi - \tfrac{1}{2}\alpha$, and so, in (3.23),

$$\tau > O \text{ when arg } (z - z_0) = \tfrac{1}{2}\pi - \tfrac{1}{2}\alpha \text{ if arg } \{-1/h''(z_0)\}^{\frac{1}{2}} = \tfrac{1}{2}\pi - \tfrac{1}{2}\alpha. \tag{3.24}$$

In this case with $h''(z_0) = a\, e^{i\alpha} = |h''(z_0)|\, e^{i\alpha}$ from (3.11), (3.23) becomes

$$z - z_0 = \left\{ \frac{2}{|h''(z_0)|} \right\}^{\frac{1}{2}} e^{i(\pi/2 - \alpha/2)}\tau + O(\tau^2),$$

$$= i\sqrt{2}\, \{h''(z_0)\}^{-\frac{1}{2}}\, \tau + O(\tau^2), \tag{3.25}$$

where here

$$\{h''(z_0)\}^{-\frac{1}{2}} = |h''(z_0)|^{-\frac{1}{2}}\, e^{-\frac{1}{2}i\alpha}. \tag{3.26}$$

With (3.25), $\tau < 0$ when z is in the A-valley where, for (3.23), arg $(z - z_0) = 3\pi/2 - \alpha/2$ in this case. As a specific example, suppose $h''(z_0) = 1$ (in a large number of cases it is just a real constant), then (3.25) simply becomes

$$z - z_0 = i\, \sqrt{2}\, \tau + O(\tau^2). \tag{3.27}$$

If the direction of integration were reversed then clearly, in place of (3.27) we would have $z - z_0 = -i\, \sqrt{2}\, \tau + O(\tau^2)$. It is best to treat each example individually. This is done in the illustrative examples in §3.2.

To complete the asymptotic problem, we require $g(z(\tau))$ as a power series, which here is

$$g(z(\tau)) = g(z_0) + (z - z_0)\, g'(z_0) + \ldots$$

$$= g(z_0) + g'(z_0) \left\{ \frac{-2}{h''(z_0)} \right\}^{\frac{1}{2}} \tau + O(\tau^2). \tag{3.28}$$

Finally (3.21) gives, with (3.23) and (3.28)

$$f(\lambda) \sim e^{\lambda h(z_0)}\, g(z_0) \left\{ \frac{-2}{h''(z_0)} \right\}^{\frac{1}{2}} \int_{-\infty}^{\infty} e^{-\lambda \tau^2}\, d\tau + \ldots \quad,$$

and so

$$f(\lambda) = g(z_0) \left\{ \frac{-2\pi}{\lambda\, h''(z_0)} \right\}^{\frac{1}{2}} e^{\lambda h(z_0)} + O\left(\frac{e^{\lambda h(z_0)}}{\lambda} \right), \tag{3.29}$$

where in line with the above discussion the specific branch of $\{-1/h''(z_0)\}^{\frac{1}{2}}$ must be chosen to be consistent with the direction of passage through the saddle-point. If the passage, for example, is from A to B in Fig. 3.2, then (3.29), with (3.26), becomes

$$f(\lambda) \sim \left(\frac{2\pi}{\lambda} \right)^{\frac{1}{2}} |h''(z_0)|^{-\frac{1}{2}}\, g(z_0)\, e^{\lambda h(z_0) + (i/2)(\pi - \alpha)}. \tag{3.30}$$

Before doing some illustrative examples in the next section, a few general comments should be made and certain points reiterated. If there is more than one saddle-point, as is very common (see the illustrative examples (ii) and (iii) in §3.2), it is naturally necessary to decide which is the relevant one for the problem: it is that which admits a possible deformation of the original contour into a path of steepest descents. If it turns out that more than one saddle-point is appropriate then that which gives the maximum ϕ is the relevant one. In this type of problem care must be taken in drawing the contour figure comparable to Fig. 3.1. Throughout we have assumed, as mentioned before, that the ends of the original contour C lie in valleys in opposite sides of the saddle-point. The method of steepest descents is often still applicable when such is not the case, as illustrated in the worked example (iv) in §3.2, since the basic idea of deforming the original contour into a path or paths of steepest descents, albeit only in one valley, is still applicable.

Various generalizations and extensions have been considered. One simple one, for example, deals with the case when $h(z)$ has a saddle-point of order n, that is the first n derivatives are zero at $z = z_0$. The procedure is essentially the same as the above but in place of (3.23), for example, we would have an $(n+1)$th root on the right-hand side and $[d^{n+1}h(z)/dz^{n+1}]_{z\,=\,z_0}$ in place of $h''(z_0)$. Others discuss, for example, the situation when $g(z)$ has a pole at the saddle-point, when two saddle-points coalesce†, when the saddle-point is close to an end point, when $\lambda h(z)$ is replaced by a general function $h(z, \lambda)$‡, and so on. Some extensions and generalizations are discussed in the books by Carrier, Krook, and Pearson (1966), Copson (1965), and Jeffreys (1962): various worked examples are also given. Research, both in methodology and in establishing rigorous proofs for various techniques in the general area is still going on.

Exercises on steepest descents are left until after §3.2 which is a section of worked examples.

3.2. Illustrative examples

Here we consider four examples which illustrate the method of steepest descents: each example demonstrates a different aspect. At every stage the general procedure of §3.1 should be kept in mind: the terminology used below is the same as that used in §3.1. The first problem is an example where there is only one saddle-point and where a preliminary transformation is required to put it in appropriate form. The second example is one in which there are two saddle-points and the correct choice of the appropriate one has to be made. This example also demonstrates the applicability of the steepest descents method to the contour integral method for obtaining solutions to a class of ordinary differential equations: this method is described. The third problem is one where there are many saddle-points. The final problem, suggested by Carrier, Krook, and Pearson (1966), demonstrates the application of the method to a case where the original contour cannot be deformed to pass through the saddle-point. Actually the leading asymptotic term in this problem can be trivially obtained by the methods of the next section, §4.1: see the footnote on equation (3.81). We discuss it by the method here for pedagogical reasons.

(i) Here we consider the Gamma function defined by (2.2) when

† Chester, C., Friedman, B. and Ursell, F. (1957). *Proc. Camb. Phil. Soc.* **53**, 599–611.
‡ Bleistein, N. and Handelsman, R. A. (1972). *J. Inst. Math. Appl.* **10**, 211–230.

the argument is *complex*. We thus consider

$$\Gamma(v+1) = \int_0^\infty e^{-w} \, w^v \, dw, \tag{3.31}$$

where the path of integration is the real axis and v is *complex* with $|\arg v| < \pi/2$. We wish to find the asymptotic expansion of $\Gamma(v+1)$ as $v \to \infty$ along some ray. The principal value of w^v is taken in (3.31) with the branch line taken from $w = 0$ to infinity along a ray lying in the domain Rl $w < 0$, the negative real axis, for example.

Integral (3.31) can be set in the appropriate form (3.1) by transforming w to z by writing

$$w = v(1+z), \quad v = \lambda e^{i\alpha}, \quad \lambda > 0, \quad |\alpha| < \frac{\pi}{2}, \tag{3.32}$$

where now z is complex and the contour, which was the positive real axis in the w-plane, becomes a contour C in the z-plane which goes from $z = -1$ ($w = 0$) to $z = \infty e^{-i\alpha}$ ($w = \infty$) as illustrated in Fig. 3.3. Integral (3.31) becomes

$$\Gamma(v+1) = v^{v+1} \, e^{-v} \int_{-1}^{\infty e^{-i\alpha}} e^{\lambda\{e^{i\alpha}[\log(1+z)-z]\}} dz, \tag{3.33}$$

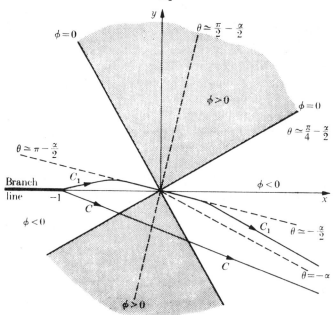

FIG. 3.3

where the principal value of the logarithm is taken and the branch line is from $z = -1$ to infinity along, say, the negative real axis or some other ray in which $\mathrm{Rl}(1+z) < 0$. We now obtain the asymptotic approximation as $\lambda \to \infty$ for the integral in (3.33). Write

$$f(\lambda) = \int_C e^{\lambda h(z)} \, dz, \tag{3.34}$$

where λ is real and positive, C is a contour from $z = -1$ to $z = \infty e^{-i\alpha}$ with $|\alpha| < \frac{1}{2}\pi$ as in Fig. 3.3, and from (3.33),

$$h(z) = e^{i\alpha}\{\log (1+z) - z\}. \tag{3.35}$$

There is a single saddle-point for $h(z)$, which is the solution of

$$h'(z) = e^{i\alpha}\left(\frac{1}{1+z} - 1\right) = 0 \Rightarrow z = 0, \tag{3.36}$$

and since $h''(0) = -1 \neq 0$, it is a saddle-point of order one. The path of steepest descents through $z = 0$ is given by

$$\psi = \operatorname{Im} h(z) = \operatorname{Im} h(0) = 0,$$

which from (3.35) for z *close* to the saddle-point $z = 0$ is, with $z = re^{i\theta}$ and $r > 0$, approximately

$$\operatorname{Im}(-e^{i\alpha} \tfrac{1}{2}z^2) \doteq 0 \Rightarrow r^2 \sin (2\theta + \alpha) \doteq 0. \tag{3.37}$$

The paths of steepest descent, and ascent, at $z = 0$ are thus tangent to the two orthogonal lines which are solutions of (3.37) $(r \neq 0)$, namely

$$\left. \begin{array}{ll} \theta = -\frac{1}{2}\alpha & \text{with its continuation } \theta = \pi - \frac{1}{2}\alpha, \\ \theta = -\frac{1}{2}\pi - \frac{1}{2}\alpha & \text{with its continuation } \theta = \frac{1}{2}\pi - \frac{1}{2}\alpha. \end{array} \right\} \tag{3.38}$$

Along the first of (3.38) and near $z = 0$

$$\phi = \mathrm{Rl}\, h(z) \doteq \mathrm{Rl}\, (-e^{i\alpha} \tfrac{1}{2}z^2) = -\tfrac{1}{2}r^2 \cos (2\theta + \alpha) < 0 = \phi_0,$$

and so the first of (3.38) is the path of steepest descents with the second being the path of steepest ascents along which $\phi > \phi_0$. Fig. 3.3, where for illustration we have taken $\frac{1}{2}\pi > \alpha > 0$, shows the mountain (shaded) and valley (unshaded) regions for the saddle-point. The contour C can now be deformed into C_1 which lies along the steepest descents path *in the vicinity of the saddle-point* and which lies completely in the valleys as indicated in Fig. 3.3.

Following (3.19) we now introduce τ by

$$h(z)-h(z_0) = e^{i\alpha}\{\log (1+z)-z\} = -\tau^2,$$

which on expanding as a Taylor series near the saddle-point $z = 0$, becomes

$$-\tfrac{1}{2}e^{i\alpha} z^2 +\ldots = -\tau^2,$$

and so for (3.23) the start of the series for $z(\tau)$, τ small, is

$$z = \pm\sqrt{2}\, e^{-\frac{1}{2}i\alpha}\, \tau+0(\tau^2),$$

where ± 1 are the two branches of $1^{\frac{1}{2}}$. On the steepest descents path $\theta = -\alpha/2$ near $z = 0$, we wish to have $\tau > 0$ and since on it $z = re^{-\frac{1}{2}i\alpha}$ we must choose the plus sign so that the $z \to \tau$ transformation is given by

$$z = \sqrt{2}\, e^{-\frac{1}{2}i\alpha}\, \tau+0(\tau^2). \tag{3.39}$$

This is consistent, as it must be, with $\tau < 0$ when z lies on the steepest descents path which, near $z = 0$, has $z = re^{i(\pi-\frac{1}{2}\alpha)}$. With (3.39), (3.34) becomes, on using (3.29),

$$f(\lambda) \sim \sqrt{2}\, e^{-\frac{1}{2}i\alpha} \int_{-\infty}^{\infty} e^{-\lambda\tau^2}\, d\tau = \left(\frac{2\pi}{\lambda}\right)^{\frac{1}{2}} e^{-\frac{1}{2}i\alpha} = \left(\frac{2\pi}{\nu}\right)^{\frac{1}{2}},$$

and so for (3.33), using (3.32),

$$\Gamma(\nu+1) \sim \sqrt{2\pi}\, e^{-\nu}\, \nu^{\nu+\frac{1}{2}} \text{ as } \nu \to \infty, \quad \left|\arg \nu\right| < \frac{\pi}{2}. \tag{3.40}$$

If we compare this result with (2.40), it is formally the same when the branch line is taken along the negative real axis, or any other ray in the left-hand half-plane. The asymptotic expression (3.40) for $\Gamma(\nu+1)$ represents the analytic continuation of $\Gamma(x+1)$, where x is real, into the complex plane where x is complex and $\left|\arg x\right| < \frac{1}{2}\pi$.

(ii) An important class of integrals, one of which was studied by Airy[†] (1838) in connection with a problem in optics, are the solutions $w(\lambda)$, in integral form, of the linear differential equation

$$w''(\lambda)- \lambda w(\lambda) = 0. \tag{3.41}$$

This equation is known as the *Airy equation* (it is usually written with a plus, rather than a minus, sign as in §6.3 below) and is of fundamental importance. It is discussed in more detail in §6.3. Solutions of it, denoted by Ai(λ), are called *Airy functions*. Equation

† Airy, G. B. (1838). *Camb. Phil. Trans.* **6**, 379–401.

(3.41) is in fact a special kind of Bessel equation (see Watson (1952)) and Airy functions are often expressed in terms of Bessel functions of fractional order.

The usual procedure for finding contour integral solutions of linear differential equations is to look for them in the specific form

$$w(\lambda) = \frac{1}{2\pi i} \int_C \phi(z) \, e^{\lambda z} \, dz, \qquad (3.42)$$

where the contour C and the complex function $\phi(z)$ are to be determined. The procedure is a kind of generalized transform method. From (3.42),

$$w''(\lambda) - \lambda w(\lambda) = \frac{1}{2\pi i} \int_C z^2 \, \phi(z) \, e^{\lambda z} \, dz - \frac{1}{2\pi i} \int_C \lambda \phi(z) \, e^{\lambda z} \, dz$$

$$= \frac{1}{2\pi i} \int_C \{z^2 \, \phi(z) + \phi'(z)\} \, e^{\lambda z} \, dz - \left[\phi \, e^{\lambda z} \right]_{z=a}^{z=b},$$

$$(3.43)$$

on integrating the second integral by parts, where $z = a$ and $z = b$ are the ends of the contour C. If we now choose the contour such that $\phi \, e^{\lambda z} \to 0$ as $z \to a$ and $z \to b$ and such that ϕ satisfies

$$\phi'(z) + z^2 \, \phi(z) = 0, \qquad (3.44)$$

then the right-hand side of (3.43) is zero and the $w(\lambda)$ given by (3.42), with such a $\phi(z)$ and contour C, is a solution of (3.41). The solution of (3.44) is $\phi = e^{-\frac{1}{3}z^3}$ and so we must choose a contour such that $\phi \, e^{\lambda z} = e^{\lambda z - \frac{1}{3}z^3} \to 0$ at its end points. If $z \to \infty$ with arg $z^3 = 0$, $2\pi, 4\pi$, for example, then $-\frac{1}{3}z^3$ is real and negative, and $e^{\lambda z - \frac{1}{3}z^3} \to 0$ as $z \to \infty$ along these rays which, in terms of arg z, are arg $z = 0$, $\frac{2}{3}\pi, \frac{4}{3}\pi$. If we choose paths, C_1, C_2 and C_3 with the end points as shown in Fig. 3.4, that is

$$\left. \begin{array}{c} C_1 \\[2mm] C_2 \\[2mm] C_3 \end{array} \right\} \text{ has end points } \begin{array}{c} \infty e^{i4\pi/3}, \ e^{i2\pi/3}, \\[2mm] \infty e^{i2\pi/3}, \ \infty, \\[2mm] \infty, \ \infty e^{i4\pi/3}, \end{array} \qquad (3.45)$$

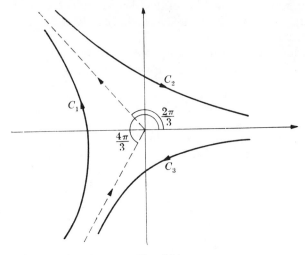

FIG. 3.4

then all three contours and allowable deformations of them are possible contours for (3.42). The contour C_1, for example, could be deformed to lie along the two straight rays $\theta = \frac{2}{3}\pi$ and $\theta = \frac{4}{3}\pi$, with the directions as indicated in Fig. 3.4. It might be thought that (3.42), with contours C_1, C_2, C_3, appears to represent three independent solutions of the differential equation (3.41) which can only have two such solutions. Of course, these three solutions are not in fact independent as is seen from the following. Those with C_2 and C_3 are dependent.

Denote by I_n the integral

$$ I_n = \frac{1}{2\pi i} \int_{C_n} e^{\lambda z - \frac{1}{3} z^3} \, dz, \quad n = 1, 2, 3, $$

with the contours C_n ($n = 1, 2, 3$) as in (3.45). Then

$$ I_1 + I_2 + I_3 = \frac{1}{2\pi i} \int_C e^{\lambda z - \frac{1}{3} z^3} dz, $$

where $C = C_1 + C_2 + C_3$ is a closed contour. Since $e^{\lambda z - \frac{1}{3} z^3}$ is analytic within C, $I_1 + I_2 + I_3 = 0$, by Cauchy's theorem. Thus I_1, I_2, and I_3 are linearly dependent.

The *Airy integral†* is defined as

$$\text{Ai}(\lambda) = \frac{1}{2\pi i} \int_C e^{\lambda z - \frac{1}{3}z^3} \, dz, \tag{3.46}$$

where λ is a given constant, complex or real, and C is any of the contours or allowable deformations of C_1, C_2, and C_3 in (3.45). We shall now use the method of steepest descents to obtain the asymptotic approximation for one of the Airy integrals (3.46) for λ real, large, and positive. Specifically we consider

$$\text{Ai}(\lambda) = \frac{1}{2\pi i} \int_{C_1} e^{\lambda Z - \frac{1}{3}Z^3} \, dZ \text{ as } \lambda \to \infty, \tag{3.47}$$

where C_1 is a contour similar to that in Fig. 3.4, with end points as given by (3.45). The integral in (3.47) is not in the appropriate form as it stands since, if $g(Z) = e^{-\frac{1}{3}Z^3}$, this dominates the integrand at the end points. We thus transform (3.47) so that it is of the form (3.1) and this is done by setting

$$Z = vz, \quad \lambda = v^2, \quad v > 0, \tag{3.48}$$

and (3.47) becomes

$$\text{Ai}(v^2) = \frac{v}{2\pi i} \int_{C_1} e^{v^3 (z - \frac{1}{3}z^3)} \, dz, \tag{3.49}$$

where C_1 in the z-plane is similar to C_1 in Fig. 3.4. Here (compare with the form (3.1))

$$h(z) = z - \tfrac{1}{3}z^3, \tag{3.50}$$

and the saddle-points are given by

$$h'(z) = 1 - z^2 = 0 \Rightarrow z = z_0 = \pm 1. \tag{3.51}$$

The question immediately arises as to which saddle-point to choose for the asymptotic analysis. We must choose the one which will allow C_1 to be deformed into the steepest descents path through it. Consider first the steepest descents paths through $z = -1$, which is given

---

† The actual integral Airy considered is that given by the real part of $\text{Ai}(\lambda)$ in (3.46) when C is $-i\infty$ to $i\infty$ and $z = iZ$, namely

$$\text{Rl} \frac{1}{2\pi} \int_{-\infty}^{\infty} e^{i(\lambda Z + \frac{1}{3}Z^3)} \, dZ = \frac{1}{\pi} \int_{0}^{\infty} \cos{(\lambda Z + \tfrac{1}{3}Z^3)} \, dZ,$$

where Z is real now and the path is the real axis.

by the appropriate curve of the two paths of

$$\psi = \text{Im } h(z) = \text{Im } h(-1) = \text{Im } (-\tfrac{2}{3}) = 0 = \psi_0.$$

With $z = x+iy$, these paths with $h(z)$ from (3.50) are

$$\text{Im } [(x+iy)\{1-\tfrac{1}{3}(x+iy)^2\}] = 0,$$

which are the curves

$$y(y^2-3x^2+3) = 0, \qquad (3.52)$$

namely, the real axis $y = 0$ and the left branch of the hyperbola $y^2 - 3x^2 + 3 = 0$ which passes through the saddle point $x = -1$, $y = 0$ as illustrated in Fig. 3.5. The asymptotes for the hyperbola we

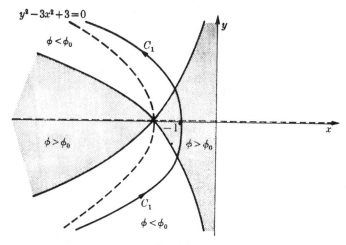

FIG. 3.5

have are the lines $y = \pm\sqrt{3}x$ or, in terms of z, $z = r\, e^{i2\pi/3}$ and $z = r\, e^{i4\pi/3}$, which for large enough r lie on C_1. If $z = -1$ is the appropriate saddle-point to take—and we shall see below that it is—then the hyperbola in Fig. 3.5 is possibly the path of steepest descents, but of course it has to be shown to be the case: this we now do. Consider $\phi = \text{Rl } h(z)$, which from (3.50) is

$$\phi = \text{Rl } [(x+iy)\{1-\tfrac{1}{3}(x+iy)^2\}] = \tfrac{1}{3}x(3y^2-x^2+3). \quad (3.53)$$

At $z = -1$, $\phi = -\tfrac{2}{3} = \phi_0$, and so on the path $y = 0$, $\phi = \tfrac{1}{3}x(3-x^2) > \phi_0$, in the vicinity of the saddle-point $x = -1$. Thus

$y = 0$ is the steepest *ascents* path. On $y^2 - 3x^2 + 3 = 0$ we see that near $z = -1$, $x \doteq -1$ and y is small, and so $x = -(1 + \frac{1}{3}y^2)^{\frac{1}{2}} \doteq -1 - \frac{1}{6}y^2 + O(y^4)$, and hence from (3.53) on the hyperbola near $z = -1$

$$\phi = -\frac{1}{3}(1 + \frac{1}{6}y^2 + \ldots)(3y^2 - 1 - \frac{1}{3}y^2 + 3)$$

$$= -\frac{2}{3} - y^2 + \ldots < -\frac{2}{3} = \phi_0.$$

Thus the hyperbolic path through $z = -1$ lies in the valleys and it is the path of steepest *descents* and further, it is the appropriate steepest descents path since the contour C_1 of (3.49) can be deformed to lie on it: see also Fig. 3.5.

Following (3.19) we now introduce the new variable τ by

$$h(z) - h(-1) = -\tau^2,$$

which from (3.50) is

$$z - \frac{1}{3}z^3 = -\frac{2}{3} - \tau^2, \tag{3.54}$$

and which on inversion gives $z(\tau)$ and hence $dz(\tau)/d\tau$ for use in the equivalent integral to (3.21). The inversion of (3.54) can, of course, be given exactly but, for our purposes, it is still simpler to consider the inversion near $z = -1$ as in (3.23) which gives, on expanding in a Taylor series and remembering that $h'(-1) = 0$,

$$z - \frac{1}{3}z^3 = -\frac{2}{3} + \frac{1}{2}[-2z]_{z=-1}(z+1)^2 + \ldots = -\frac{2}{3} - \tau^2.$$

That is

$$(z+1)^2 = -\tau^2 + \ldots \Rightarrow z+1 = \pm i\,\tau + 0(\tau^2). \tag{3.55}$$

When z is on the upper branch of the hyperbolic steepest descents path we have, near $z = -1$, $\arg z \doteq \pi$ and so $\arg(z+1) \doteq \pi/2$, which shows that the argument of the right-hand side of (3.55) is $\pi/2$. Thus since τ is real and $\tau > 0$ here we must choose $+i$ as the appropriate branch of $(-1)^{\frac{1}{2}}$ which gives, from (3.55), the correct transformation as $z+1 \doteq i\tau$ and hence $dz/d\tau \doteq i$. Note again that $\tau < 0$ is consistent with z on the lower branch of the hyperbola near $z = -1$, since here $\arg(z+1) \doteq -\pi/2$.

On deforming C_1 to lie along the hyperbola through $z = -1$, equation (3.49) becomes, in terms of τ,

$$\text{Ai}\,(v^2) \sim \frac{v}{2\pi i}\,e^{-\frac{2}{3}v^3}\int_{-\infty}^{\infty} e^{-v^3\tau^2}\,i\,d\tau + \dots \text{ as } v \to \infty,$$

and since $\int_{-\infty}^{\infty} e^{-v^3\tau^2}\,d\tau = \left(\frac{\pi}{v^3}\right)^{\frac{1}{2}}$,

$$\text{Ai}\,(v^2) \sim \tfrac{1}{2}(\pi v)^{-\frac{1}{2}}\,e^{-\frac{2}{3}v^3} + \dots \quad .$$

From (3.48), $v^2 = \lambda$ and so Ai (λ), defined by (3.47), is

$$\text{Ai}\,(\lambda) \sim \tfrac{1}{2}\pi^{-\frac{1}{2}}\lambda^{-\frac{1}{4}}\,e^{-\frac{2}{3}\lambda^{3/2}} + \dots \text{ as } \lambda \to \infty. \tag{3.56}$$

An analysis of the next term shows it to be $O(\lambda^{-\frac{1}{4}}\lambda^{-\frac{3}{2}}e^{-\frac{2}{3}\lambda^{3/2}})$.

To complete the analysis of this example, we return to the question of the other saddle-point at $z = 1$. Since $h(1) = \frac{2}{3}$, Im $h(1) = 0$, and so the paths $\psi = \text{Im } h(z) = 0$ are given by the same expression as before, namely (3.52), but now we have the *right-hand* branch of the hyperbola $y^2 - 3x^2 + 3 = 0$, since it passes through the saddle-point $z = 1$, together with the real axis $y = 0$ as before. On this branch of the hyperbola near $z = 1$, $x \doteq 1$ and y is small so from (3.53)

$$\phi = \tfrac{1}{3}x(3y^2 - x^2 + 3) > \phi_0 = \tfrac{2}{3} = \text{Rl } h(1),$$

and so this time the hyperbola is the steepest *ascents* path and is in the mountain regions as indicated in Fig. 3.6. On the other $\psi = \psi_0$ path, the real axis $y = 0$, $\phi < \phi_0$ near to $z = 1$, which makes it a steepest descents path. The mountain-valley regions in this case are illustrated in Fig. 3.6, from which it is clear that C_1 cannot be deformed into the real axis, which is the steepest descents path. Thus $z = 1$ is *not* a possible choice of saddle-point for Ai(λ) in (3.46) where C is the contour C_1 of (3.45).

If we consider Ai(λ) in (3.46) with contours C_2 and C_3 of (3.45), then it is clear from the above discussion that the appropriate saddle-point to take in their case *is* $z = 1$: each contour (see C_2 in Fig. 3.6, for example) can be deformed to pass through $z = 1$ with the final contour lying entirely in the valley regions $\phi < \phi_0$. An examination of these integrals with contours C_2 and C_3 is left as an exercise (exercise 3 below).

From the differential equation (3.41), with $\lambda > 0$, heuristically we would expect the solutions to display some kind of *exponential* behaviour, and (3.56) confirms this (see also exercise 3). On the other hand, if $\lambda < 0$ in the differential equation (3.41), heuristically

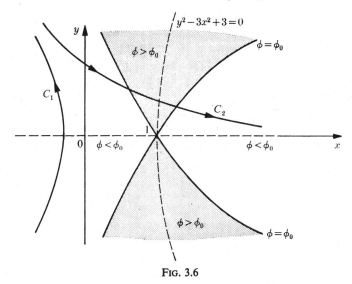

FIG. 3.6

we would expect the solutions to have an *oscillatory*† nature. This is exactly the case and it is found (exercise 4) that

$$\text{Ai}(-\lambda) = \frac{1}{2\pi i} \int_{C_1} e^{-\lambda z - \frac{1}{3}z^3} \, dz$$

$$\sim \frac{1}{\sqrt{\pi}\,\lambda^{\frac{1}{4}}} \sin\left(\tfrac{2}{3}\lambda^{\frac{3}{2}} + \pi/4\right) \text{ as } \lambda \to \infty. \qquad (3.57)$$

Thus with the same contour C_1 as in Fig. 3.4, although $\text{Ai}(\lambda) \to 0$ as $\lambda \to \pm\infty$, the manner in which it does so is exponential in one case, as $\lambda \to \infty$ (see (3.56)), and algebraic and oscillatory in the other, as $\lambda \to -\infty$ (see (3.57)). The Airy function denoted by $\text{Bi}(\lambda)$, which is another solution, independent of $\text{Ai}(\lambda)$, of the Airy equation (3.41), is defined in exercise 3 where its asymptotic form as $\lambda \to \infty$ is given: exercise 4 gives its asymptotic expansion as $\lambda \to -\infty$. These results are of fundamental importance as seen below in §6.3.

(iii) Bessel functions of one kind or another appear in a remarkably large number of practical problems and are the most widely studied

---

† This is, of course, also an exponential behaviour. When we use the words 'exponential' and 'oscillatory' we simply want to emphasize the fact that the function has an exponential behaviour with a purely real and purely imaginary exponent respectively.

special functions. Their asymptotic expansions have been considered in some detail. The main reference book on them is by Watson (1952), who obtains asymptotic forms for a variety of Bessel functions in various situations. As mentioned above, the Airy integral is a special type of Bessel function. In view of the general importance of Bessel functions, in this example we consider a particular type of Bessel function which is known as a *Hankel* function: it also illustrates a case where there are many saddle-points.

We consider the *Hankel function* of the first kind, $H_s^{(1)}(\lambda)$, which

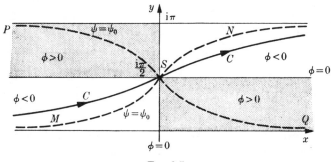

FIG. 3.7

is defined by

$$H_s^{(1)}(\lambda) = \frac{1}{\pi i} \int_C e^{\lambda \sinh z - sz}\, dz, \tag{3.58}$$

where s, the order of the Hankel function, is a real positive number, not necessarily an integer, and C is a contour which starts at $-\infty$ and ends at $\infty + \pi i$ as in Fig. 3.7. We shall look for the asymptotic expansion for λ real and large[†] and for any given s. Actually the asymptotic approximation we shall obtain is of practical use only if $\lambda \gg s$: a case in which $\lambda = O(s)$ is left as exercise 6. For (3.58) we have, on comparison with (3.1),

$$h(z) = \sinh z, \tag{3.59}$$

and so the saddle-points are given by

$$h'(z) = \cosh z = 0, \quad z = z_0 = \frac{2n+1}{2}\pi i, \quad n = 0, \pm 1, \pm 2 \ldots, \tag{3.60}$$

[†] As $\lambda \to 0$, on the other hand, $H_s^{(1)}(\lambda)$ is singular (see Watson (1952), for example).

and since

$$h''(z_0) = \sinh \frac{2n+1}{2} \pi i \neq 0,$$

represents an infinite number of first order saddle-points. In view of the contour we have in (3.58), we expect the relevant saddle-point to be that at $z_0 = i\pi/2$, which we now consider. It will become clear, and it will be pointed out at the appropriate place, that the other saddle-points are not suitable for the asymptotic evaluation of (3.58) with the specific given contour C.

The steepest descents path through the saddle-point $z_0 = i\pi/2$ is, from (3.59), given by the appropriate curve

$$\psi = \text{Im } (z) = \text{Im sinh } z = \text{Im } h(z_0) = \text{Im sinh } \frac{i\pi}{2} = 1 = \psi_0,$$
(3.61)

which, with $z = x + iy$, becomes

$$\cosh x \sin y = 1. \tag{3.62}$$

As usual two curves are obtained since in (3.62) for a given x, there are two ys, one in the 0 to $\pi/2$ range and another in the $\pi/2$ to π range. These two curves, from (3.62), are the dashed lines MSN and PSQ in Fig. 3.7: they are orthogonal at the saddle-point S.

Proceeding as before, we now calculate $\phi = \text{Rl } h(z)$ along these curves to determine which is the steepest descents path. Since, from (3.59),

$$\text{Rl } h(z) = \phi = \sinh x \cos y \Rightarrow \text{Rl } h(z_0) = \phi_0 = 0; \quad (3.63)$$

if we move away from $z = i\pi/2$ along the line SQ in Fig. 3.7, $x > 0$, $0 < y < \pi/2$ and so, from (3.63), $\phi > \phi_0$. Similarly, along SP $x < 0$, $\pi/2 < y < \pi$ and again $\phi > \phi_0$. This shows that PSQ is the path of steepest ascent and is in the mountain domain. If we now leave S along the paths SN and SM we have $x > 0$, $\pi/2 < y < \pi$ and $x < 0$, $0 < y < \pi/2$, respectively, and so from (3.63) $\phi < \phi_0 = 0$. Thus the path through MSN is the path of steepest descents, which lies in the valley regions. The two lines $x = 0$ and $y = \pi/2$ give $\phi = \phi_0 = 0$ from (3.63), and so the valley-mountain domains in which $\phi < 0$ and $\phi > 0$ (shaded), respectively, are as illustrated in Fig. 3.7. It is now clear that the original contour C can be deformed to lie along the steepest descents path MSN, and so for the asymptotic evaluation of the integral (3.58) we now have

$$H_s^{(1)}(\lambda) = \frac{1}{\pi i} \int_{-\infty}^{\infty+\pi i} e^{\lambda \sinh z} \, e^{-sz} \, dz, \qquad (3.64)$$

where the integration path is the steepest descents one along *MSN*. From the above we see that the contour *C* could not be deformed into the steepest descents path through any saddle-point other than the one $z_0 = i\pi/2$.

It is convenient to change the variable so that the saddle-point is the origin, in which case we write,

$$w = z - \frac{i\pi}{2}, \qquad (3.65)$$

and (3.64) becomes

$$H_s^{(1)}(\lambda) = \frac{1}{\pi i} \int_{-\infty-i\pi/2}^{\infty+i\pi/2} e^{i\lambda \cosh w} \, e^{-sw} \, e^{-is\pi/2} \, dw, \qquad (3.66)$$

where the integration is along the steepest descents path. In the now usual way we introduce the new variable τ, following (3.19), by writing

$$i \cosh w - i = -\tau^2. \qquad (3.67)$$

On the path of steepest descents τ is real, of course. We now expand (3.67) for *w* small and on the path of steepest descents we have, from (3.67),

$$\tfrac{1}{2} i \, w^2 + \ldots = -\tau^2,$$

and so

$$w = \sqrt{2} \, i^{\frac{1}{2}} \, \tau + O(\tau^2). \qquad (3.68)$$

We must now decide which branch of $i^{\frac{1}{2}}$, namely, $e^{i\pi/4}$ or $e^{-i\pi/4}$, is the appropriate one. Because of the direction of integration through the saddle-point we must have $\tau > 0$ on the steepest descents path where Rl $w > 0$, $\pi/2 >$ Im $w > 0$ (on comparison with Fig. 3.7, that is $x > 0$, $\pi/2 < y < \pi$), which means $0 < \arg w < \pi/2$ there. Thus we must choose $e^{i\pi/4}$ for $i^{\frac{1}{2}}$, and so for (3.68) we take $w = \sqrt{2} \, e^{i\pi/4} \, \tau + O(\tau^2)$ and hence $dw/d\tau = \sqrt{2} \, e^{i\pi/4} + O(\tau)$ on the integration path for (3.66). With (3.68), for τ small.

$$g(w(\tau)) = e^{-sw} = 1 + O(\tau),$$

and so, finally, (3.66) with (3.67), and $w = \sqrt{2} \, e^{i\pi/4} \, \tau + \ldots$ as the appropriate form of (3.68), and $\lambda \to \infty$,

$$H_s^{(1)}(\lambda) \sim \frac{1}{\pi i} \int_{-\infty}^{\infty} e^{i\lambda} e^{-\lambda \tau^2} e^{-is\pi/2} e^{i\pi/4} \sqrt{2} \, d\tau + \dots \,,$$

which gives for (3.58)

$$H_s^{(1)}(\lambda) \sim e^{i(\lambda - s\pi/2 - \pi/4)} \left(\frac{2}{\lambda \pi}\right)^{\frac{1}{2}} + \dots \text{ as } \lambda \to \infty. \tag{3.69}$$

The Hankel function of the second kind is given by $H_s^{(2)}(\lambda) = \overline{H_s^{(1)}(\lambda)}$, the complex conjugate of $H_s^{(1)}(\lambda)$, and so from (3.69) we have, immediately,

$$H_s^{(2)}(\lambda) \sim e^{-i(\lambda - s\pi/2 - \pi/4)} \left(\frac{2}{\lambda \pi}\right)^{\frac{1}{2}} + \dots \text{ as } \lambda \to \infty. \tag{3.70}$$

The Bessel function of the first kind, $J_s(\lambda)$, may be defined (see, for example, Watson (1958)) by

$$J_s(\lambda) = \tfrac{1}{2}\{H_s^{(1)}(\lambda) + H_s^{(2)}(\lambda)\},$$

and so from (3.69) and (3.70) we get the asymptotic approximation to the Bessel function $J_s(\lambda)$ as

$$J_s(\lambda) \sim \left(\frac{2}{\lambda \pi}\right)^{\frac{1}{2}} \cos\left(\lambda - s\frac{\pi}{2} - \frac{\pi}{4}\right) + \dots \text{ as } \lambda \to \infty. \tag{3.71}$$

(iv) As a final example, we obtain the asymptotic approximation as $\lambda \to \infty$ of the integral

$$f(\lambda) = \int_0^1 z^{-\frac{1}{2}} e^{i\lambda(z + z^2)} \, dz, \tag{3.72}$$

where the path of integration is along the real axis. The principal value of $z^{-\frac{1}{2}}$ is taken and the branch line drawn along the negative real axis. Here $h(z) = i(z + z^2)$ and there is a single saddle-point of order one given by

$$h'(z) = i(1 + 2z) = 0 \Rightarrow z = -\tfrac{1}{2}. \tag{3.73}$$

The steepest descents path through $z = -\tfrac{1}{2}$ is given by

$$\psi = \text{Im } i(z + z^2) = \text{Im } h(-\tfrac{1}{2}) = \text{Im } (-i/4) = -\tfrac{1}{4} = \psi_0,$$

which, with $z = x+iy$, are the two straight lines

$$(x+\tfrac{1}{2})^2 - y^2 = 0. \tag{3.74}$$

Here

$$\phi = \text{Rl } i(z+z^2) = -y(1+2x),$$

and so $\phi = 0 = \phi_0$ at $z = -\tfrac{1}{2}$. Figure 3.8 illustrates the valley and mountain (shaded) regions in which $\phi < 0$ and $\phi > 0$ (shaded). The path of steepest descent through the saddle point $z = -\tfrac{1}{2}$ is thus, from (3.74), the line $x+\tfrac{1}{2} = y$. However, in this problem, the original contour from 0 to 1 on the real axis *cannot* be deformed to lie along this steepest descents path and pass through the saddle-

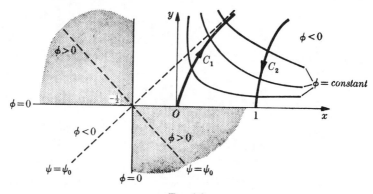

FIG. 3.8

point. If we recall the general theory and concepts of §3.1, however, we can see how to use the principles of the method of steepest descents on the problem. Remember that an appropriate path must be along a ψ-line to exclude oscillations and to ensure that ϕ changes as rapidly as possible along it. Here we have *two* steepest descents paths, one from each of $z = 0$ and $z = 1$, both of which start on the $\phi = 0$ line (the x-axis here) and tend to the steepest descents path $x+\tfrac{1}{2} = y$, through the saddle-point, for large positive x and y: these are denoted by C_1 and C_2 in Fig. 3.8. On C_1 and C_2, ϕ has its maximum value at $z = 0$ and $z = 1$ respectively. These paths, which are steepest descent ones, can now be used for the asymptotic evaluation of (3.72). These two paths, C_1 and C_2, are given by the appropriate parts of the two hyperbolae

$$\psi = \text{Im } i(z+z^2) = \textit{constant}$$

through $z = 0$ and $z = 1$ which are respectively

$$
\begin{rcases}
\text{Im } i(z+z^2) = 0, \text{ namely, } x^2 + x - y^2 = 0, \\[2mm]
\text{Im } i(z+z^2) = 2, \text{ namely, } x^2 + x - y^2 = 2,
\end{rcases}
\tag{3.75}
$$

and

both of which have asymptotes (3.74). The original contour from 0 to 1 on the real axis *can* be deformed into the two upper branches C_1 and C_2 of these hyperbolae (3.75), which lie in the *same* valley, and so $f(\lambda)$, from (3.72), may be written as

$$
f(\lambda) = \left(\int_{C_1} + \int_{C_2} \right) z^{-\frac{1}{2}} e^{i\lambda(z+z^2)} \, dz,
\tag{3.76}
$$

where C_1 is the steepest descent path, the first of (3.75), from $z = 0$ to $z = \infty e^{i\pi/4}$ and C_2 is the steepest descent path, the second of (3.75), from $\infty e^{i\pi/4}$ to 1. Note the direction in C_2. Since C_1 is Im $h(z) = 0$ and C_2 is Im $h(z) = 2$, if we introduce the new variables τ and η by

$$
\begin{rcases}
i\tau = z+z^2 \Rightarrow z = -\tfrac{1}{2} + \tfrac{1}{2}(1+4i\tau)^{\frac{1}{2}} \quad \text{for } C_1 \\[2mm]
2+i\eta = z+z^2 \Rightarrow z = -\tfrac{1}{2} + \tfrac{1}{2}(9+4i\eta)^{\frac{1}{2}} \text{ for } C_2,
\end{rcases}
\tag{3.77}
$$

then τ and η are *real* when z lies respectively on the steepest descent contours C_1 and C_2. In (3.77) the specific roots chosen for z are those giving $z = 0$ when $\tau = 0$ in the first equation and $z = 1$ when $\eta = 0$ in the second. With (3.77), (3.76) now becomes

$$
f(\lambda) = \int_0^\infty e^{-\lambda\tau} \{z(\tau)\}^{-\frac{1}{2}} \frac{dz(\tau)}{d\tau} \, d\tau -
$$

$$
- \int_0^\infty e^{2i\lambda} e^{-\lambda\eta} \{z(\eta)\}^{-\frac{1}{2}} \frac{dz(\eta)}{d\eta} \, d\eta.
\tag{3.78}
$$

Since $z^{-\frac{1}{2}} \, dz/d\tau$ and $z^{-\frac{1}{2}} \, dz/d\eta$ can be obtained simply as a Taylor series in τ and η respectively from (3.77), in this case we can get the complete asymptotic expansion as $\lambda \to \infty$ using Watson's lemma from §2.1. Here we shall obtain the leading term from each integral in (3.78). From (3.77), for τ small,

$$
z^{-\frac{1}{2}} \frac{dz}{d\tau} = \left\{ -\tfrac{1}{2} + \tfrac{1}{2}(1+4i\tau)^{\frac{1}{2}} \right\}^{-\frac{1}{2}} i(1+4i\tau)^{-\frac{1}{2}}
$$

$$
= i^{\frac{1}{2}} \tau^{-\frac{1}{2}} + O(\tau^{\frac{1}{2}}),
\tag{3.79}
$$

and, since on C_1 for z small $0 < \arg z < \pi/2$, we must choose the branch $i^{\frac{1}{2}} = e^{i\pi/4}$. Also, from (3.77), for η small,

$$z^{-\frac{1}{2}}\frac{dz}{d\eta} = \{-\tfrac{1}{2}+\tfrac{1}{2}(9+4i\eta)^{\frac{1}{2}}\}^{-\frac{1}{2}}\, i(9+4i\eta)^{-\frac{1}{2}}$$

$$= \frac{i}{3}+O(\eta). \tag{3.80}$$

With (3.79) and (3.80), the asymptotic approximation for $f(\lambda)$ from (3.78) is given by

$$f(\lambda) \sim \int_0^\infty e^{-\lambda\tau}\{e^{i\pi/4}\tau^{-\frac{1}{2}}+O(\tau^{\frac{1}{2}})\}\, d\tau - \frac{i}{3}e^{2i\lambda}\int_0^\infty e^{-\lambda\eta}[1+O(\eta)]\, d\eta,$$

and so, since $\int_0^\infty \tau^{-\frac{1}{2}}\, e^{-\lambda\tau}\, d\tau = \lambda^{-\frac{1}{2}}\, \Gamma(\tfrac{1}{2}) = (\pi/\lambda)^{\frac{1}{2}}$,

$$f(\lambda) = \int_0^1 z^{-\frac{1}{2}} e^{i\lambda(z+z^2)}\, dz \sim \left(\frac{\pi}{\lambda}\right)^{\frac{1}{2}} e^{i\pi/4} - \frac{i}{3\lambda}e^{2i\lambda}+O(\lambda^{-\frac{3}{2}}) \text{ as } \lambda \to \infty.$$
$$\tag{3.81}\dagger$$

Exercises

1. Using the method of steepest descents, discuss and show that the asymptotic approximations for the following are as given.

   (i) Hankel's integral (which is related to the Gamma function), defined by

   $$\frac{1}{\Gamma(\lambda)} = \frac{1}{2\pi i}\int_C e^z\, z^{-\lambda}\, dz \sim e^\lambda\, \lambda^{-\lambda}\left(\frac{\lambda}{2\pi}\right)^{\frac{1}{2}} \text{ as } \lambda \to \infty$$

   for real λ and C the contour from $-\infty$ round the origin as indicated in Fig. 3.9 and then to $-\infty$ again.

† The first term here can be very simply obtained by using the results of the next section, §4.1. If we write $z = w^2/\lambda^{\frac{1}{2}}$,

$$\int_0^1 z^{-\frac{1}{2}} e^{i\lambda(z+z^2)}\, dz = \frac{2}{\lambda^{\frac{1}{4}}}\int_0^\infty e^{iw^4} e^{i\lambda^{\frac{1}{2}}w^2}\, dw \sim \frac{1}{\lambda^{\frac{1}{4}}}\left(\frac{\pi}{\lambda^{\frac{1}{2}}}\right)^{\frac{1}{2}} e^{i\pi/4}$$

on using the result (4.12): it is half the value there because the range of integration here is from 0 to ∞ rather than $-\infty$ to ∞.

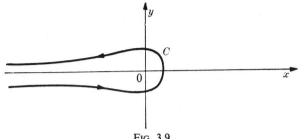

FIG. 3.9

(ii) $f(\lambda) = \int\limits_{-\infty}^{\infty} e^{i\lambda z} (1+z)^{-\lambda} \, dz \sim \left(\frac{2\pi}{\lambda}\right)^{\frac{1}{2}} e^{\lambda} e^{i\lambda\pi/2}$,

for λ real and large, where the path of integration is the real axis, and the branch line is a ray from $z = -1$ to infinity

in $\dfrac{\pi}{2} < \arg (1+z) < \pi$.

2. If

$$f(\lambda, t) = \frac{1}{2\pi i} \int\limits_{c-i\infty}^{c+i\infty} z^{-1} \, e^{\lambda(tz - z^{1/2})} \, dz,$$

where c, t and λ are real and positive and the principal value of $z^{\frac{1}{2}}$ is taken with the branch line along the negative real axis, show, by the method of steepest descents, that

$$f(\lambda, t) \sim 2 \left(\frac{t}{\pi\lambda}\right)^{\frac{1}{2}} e^{-\lambda/4t} \text{ as } \lambda \to \infty.$$

3. Discuss the asymptotic evaluation and obtain the leading term in the asymptotic expansion of the Airy integral

$$\text{Ai}(\lambda) = \frac{1}{2\pi i} \int\limits_{C} e^{\lambda z - \frac{1}{3} z^3} \, dz \text{ as } \lambda \to \infty$$

for λ real and where the contour C is from $\infty \, e^{i2\pi/3}$ to ∞. namely C_2 in Fig. 3.4 above. Using this result and that in the above section, find the leading term in the asymptotic expansion of $\text{Ai}(\lambda)$ when C is C_3 in Fig. 3.4, that is from ∞ to $\infty \, e^{i4\pi/3}$. Hence show that $\text{Bi}(\lambda)$, another solution of the Airy equation (3.41), defined by

$$\text{Bi}(\lambda) = \frac{1}{2\pi} \left(\int\limits_{C_2} - \int\limits_{C_3}\right) e^{\lambda z - \frac{1}{3} z^3} \, dz$$
$$\sim \pi^{-\frac{1}{2}} \lambda^{-\frac{1}{4}} e^{\frac{2}{3}\lambda^{3/2}} \text{ as } \lambda \to \infty.$$

4. Consider the asymptotic evaluation of the Airy integral with *negative* argument, namely

$$\text{Ai}(-\lambda) = \frac{1}{2\pi i} \int\limits_{C_1} e^{-\lambda z - \frac{1}{3} z^3} \, dz,$$

where C_1 is the contour in Fig. 3.4. This function is a solution of the Airy equation $w'' + \lambda w = 0$. First show, by writing $Z = \lambda^{\frac{1}{3}} z$, that the saddle-points in the Z-plane are at $Z = \pm i$, that the contour deforms, for the steepest descents evaluation, into the two contours as shown in

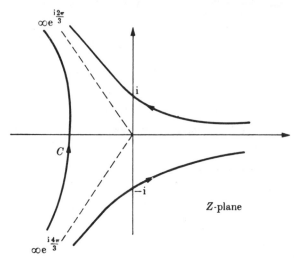

$$\infty e^{i\frac{2\pi}{3}}$$

$$i$$

$$C$$

$$-i$$

Z-plane

$$\infty e^{i\frac{4\pi}{3}}$$

FIG. 3.10

Fig. 3.10, and that the contributions from the saddle-points at $\mp i$ are

$$\frac{1}{2i\sqrt{\pi}}\, \lambda^{-\frac{1}{4}}\, e^{\pm i(\frac{2}{3}\lambda^{3/2} + \pi/4)},$$

which together give the asymptotic expansion as

$$\text{Ai}(-\lambda) \sim \frac{1}{\lambda^{\frac{1}{4}}\sqrt{\pi}}\, \sin\left(\tfrac{2}{3}\lambda^{\frac{3}{2}} + \frac{\pi}{4}\right) \quad \text{as } \lambda \to \infty.$$

With the definition of $\text{Bi}(\lambda)$ given in question 3 and this result for $\text{Ai}(-\lambda)$, show that

$$\text{Bi}(-\lambda) \sim \frac{1}{\lambda^{\frac{1}{4}}\sqrt{\pi}}\, \cos\left(\tfrac{2}{3}\lambda^{\frac{3}{2}} + \frac{\pi}{4}\right) \quad \text{as } \lambda \to \infty.$$

(Note that $\text{Ai}(-\lambda)$ and $\text{Bi}(-\lambda)$ are *oscillatory* and tend to zero *algebraically* like $O(\lambda^{-\frac{1}{4}})$. This should be compared with the *exponential* decay of $\text{Ai}(\lambda)$ and the *exponential* growth of $\text{Bi}(\lambda)$ as $\lambda \to \infty$: by 'oscillatory' and 'exponential' we mean an exponential behaviour with an imaginary and real exponent respectively.)

5. Find the asymptotic approximation for

$$f(\lambda, s) = \frac{1}{\pi i} \int_C e^{\lambda \sinh z - sz}\, dz \quad \text{as } \lambda \to \infty$$

for λ and s real and positive and where the contour C, as indicated in Fig. 3.11 starts at $-\infty + n\pi i$ and ends at $\infty + (n+1)\pi i$, where n is an integer.

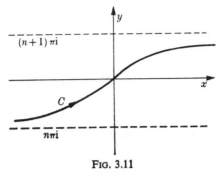

FIG. 3.11

6. Find the asymptotic expansion of the Hankel function

$$H_{c\lambda}^{(1)}(\lambda) = \frac{1}{\pi i} \int_{-\infty}^{\infty+\pi i} e^{\lambda(\sinh z - cz)} \, dz,$$

where $c > 1$ is a real positive constant. (It is algebraically convenient to write $c = \cosh a$.)

4

Method of stationary phase

4.1. Method of stationary phase

THE integrals for which this method was originally developed by
Stokes† and Kelvin‡ are of the general form

$$f(\lambda) = \int_a^b g(t)\, e^{i\lambda h(t)}\, dt, \qquad (4.1)$$

where a, b, $g(t)$, $h(t)$, t, and λ are all real. Asymptotic expansions are
sought as $\lambda \to \infty$. It should be said here that if $g(t)$ and $h(t)$ can be
suitably analytically continued off the real axis then the class of
integrals (4.1) can be treated, as discussed briefly below, by the
method of steepest descents in §3.1. However, the original method of
stationary phase antecedes the steepest descents one. There are three
main reasons for considering it separately. First, the physical idea
and motivation behind the original exposition of the method are
interesting and instructive. Secondly, we would like to be able to
deal with such integrals by considering the integration along the real
axis only. Thirdly, the fact that integrals like (4.1) play such a fundamental role in the study of general wave motion, as will be seen in
§4.2 below, is in itself a valid reason for obtaining the asymptotic
expansion. In §4.2 a brief introduction is given to dispersive wave
motion which is of current interest and practical importance: the
most exciting developments in the subject have appeared since about
1960.

In this section we briefly discuss the physical idea behind the
method, which is quite different from that behind the method of

---

† Stokes, G. G. (1857). *Camb. Phil. Trans.* **10**, 106–128. (Also in, *Mathematical
and physical papers*. (1883) Cambridge University Press.)
‡ Lord Kelvin (W. Thomson) (1887). *Phil. Mag.* **23**, 252–255. (Also in *Mathematical and physical papers*. (1910) Cambridge University Press.)

steepest descents, and develop the procedure for obtaining the asymptotic approximation for $f(\lambda)$ as $\lambda \to \infty$†. Before doing this, let us first see how the class of integrals (4.1) can, in the appropriate circumstances, be treated by the method of steepest descents. If $g(t)$ and $h(t)$ in (4.1), defined on $a \leqq t \leqq b$, can be continued analytically into a larger domain which includes that part of the real axis $a \leqq$ Rl $t \leqq b$, then we are, in effect, considering integrals like (3.1) in which $\lambda = i|\lambda|$ is purely imaginary. In place of (3.1) we then have integrals like

$$\int_C g(z)\, e^{|\lambda|ih(z)}\, dz, \tag{4.2}$$

and so the path of steepest descents for (4.2), with $h(z) = \phi + i\psi$ from (3.5) as before, is obtained from

$$\text{Im } ih(z) = \phi = constant, \tag{4.3}$$

which means that the path is along a line of constant altitude in the sense of §3.1.

Let us now consider the integral (4.1) with the conditions given there. The term $e^{i\lambda h(t)}$ is purely oscillatory, and so we cannot exploit exponential decay of the integrand from some maximum as was done, for example, in Watson's lemma (§2.1) and Laplace's method (§2.2). However, if λ is large, the oscillations of $e^{i\lambda h(t)}$ are very dense, in the sense that many oscillations occur in a very small range of t, and so for λ large enough there is the possibility of exploiting the cancellation properties of the positive and negative parts of the oscillations almost everywhere. To see the practical application of this idea, let us consider by way of illustration the real part of (4.1) with $g(t) = 1$. We thus examine

$$F(\lambda) = \int_a^b \cos \lambda h(t)\, dt, \tag{4.4}$$

in which a, b, λ and $h(t)$ are real. Let us consider $h(t)$ to have a single turning point which is a simple relative maximum or minimum, at $t = t_0$ where $a < t_0 < b$ and where

$$h'(t_0) = 0, \quad h''(t_0) \neq 0. \tag{4.5}$$

Practically, if $h(t)$ has more than one turning point in the given range, we simply split the interval up into sub-intervals in each of which

† By the Riemann-Lebesque theorem $f(\infty) = 0$, of course. Here we are interested in the *manner* in which $f(\lambda) \to 0$ as $\lambda \to \infty$.

$h(t)$ has only one turning point. Figure 4.1 shows a typical $h(t)$, where for illustration, we have taken $h(t_0)$ to be a maximum. In Fig. 4.1(a) and Fig. 4.1(b) the effect on the integrand of increasing λ is illustrated: the λ in Fig. 4.1(b) is very much larger than that in Fig. 4.1(a).

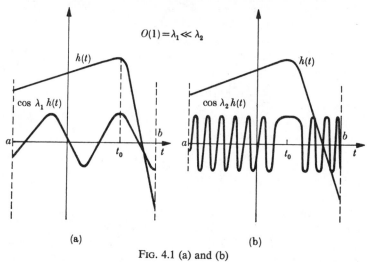

FIG. 4.1 (a) and (b)

As $\lambda \to \infty$ the oscillations are such that, in the integration of $\cos \lambda h(t)$, the positive and negative parts effectively cancel each other out, except in the vicinity of $t = t_0$, where $h'(t_0) = 0$, and possibly in the neighbourhoods of $t = a$ and $t = b$: we come back to these end points later. Near $t = t_0$ the *phase* $\lambda h(t)$ is approximately constant or *stationary*: for $t - t_0$ small $h(t) - h(t_0) = O((t-t_0)^2)$, whereas in the vicinity of any other point, τ say, $h(t) - h(\tau) = O(t - \tau)$. In the vicinity of t_0, therefore, the oscillations will *not* cancel each other out: the type of situation which obtains there is illustrated in Fig. 4.1(b). Thus the integral $F(\lambda)$ in (4.4) as $\lambda \to \infty$ is *asymptotically equal* to the integral of $\cos \lambda h(t)$ only over a small δ-neighbourhood of t_0, and is given by

$$F(\lambda) = \mathrm{Rl} \int_a^b e^{i\lambda h(t)} \, dt$$

$$\sim \mathrm{Rl} \int_{t_0-\delta}^{t_0+\delta} e^{i\lambda h(t)} \, dt. \tag{4.6}$$

Near $t = t_0$,

$$h(t) = h(t_0) + \tfrac{1}{2} h''(t_0) \, (t - t_0)^2 + \ldots, \tag{4.7}$$

which suggests (compare with §2.2) introducing a new variable s by

$$h(t) - h(t_0) = \pm s^2, \qquad (4.8)$$

where $+s^2$ is used when $h''(t_0) > 0$ and $-s^2$ when $h''(t_0) < 0$. For t in the neighbourhood of t_0, we can invert (4.8) using (4.7) to get t in terms of s as (compare with (2.27))

$$t - t_0 = \left\{ \frac{2}{|h''(t_0)|} \right\}^{\frac{1}{2}} s + O(s^2), \qquad (4.9)$$

which covers both cases: $h''(t_0) > 0$ and $h''(t_0) < 0$. The asymptotic approximation (4.6), using (4.9), now becomes

$$F(\lambda) \sim \mathrm{Rl} \left\{ \frac{2}{|h''(t_0)|} \right\}^{\frac{1}{2}} e^{\mathrm{i}\lambda h(t_0)} \int_{-s_1}^{s_1} e^{\pm\mathrm{i}\lambda s^2} \{1 + O(s)\} \, \mathrm{d}s, \qquad (4.10)$$

where we have set

$$s_1 = \delta \left\{ \frac{2}{|h''(t_0)|} \right\}^{-\frac{1}{2}}.$$

If we now make the change of variable $\eta = s\sqrt{\lambda}$, the last integral becomes, asymptotically as $\lambda \to \infty$,

$$F(\lambda) \sim \mathrm{Rl} \left\{ \frac{2}{\lambda|h''(t_0)|} \right\}^{\frac{1}{2}} e^{\mathrm{i}\lambda h(t_0)} \int_{-\infty}^{\infty} e^{\pm\mathrm{i}\eta^2} \left\{ 1 + \frac{1}{\sqrt{\lambda}} O(\eta) \right\} \, \mathrm{d}\eta\dagger$$

† To evaluate this integral consider $\int_C e^{-z^2} \, \mathrm{d}z$, where C is the closed contour shown in Fig. 4.2. Since e^{-z^2} is analytic in the domain enclosed by C, we have, by Cauchy's theorem,

$$\int_C e^{-z^2} \, \mathrm{d}z = 0 = \int_0^R e^{-r^2} \, \mathrm{d}r + \mathrm{i}R \int_0^{\pi/4} e^{-R^2 e^{2\mathrm{i}\theta}} e^{\mathrm{i}\theta} \, \mathrm{d}\theta - e^{\mathrm{i}\pi/4} \int_0^R e^{-\mathrm{i}r^2} \, \mathrm{d}r.$$

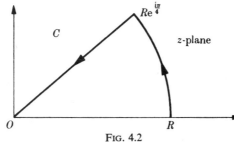

FIG. 4.2

As $R \to \infty$ the second integral on the right-hand side tends to zero, and we are left with

$$\int_0^{\infty} e^{-\mathrm{i}r^2} \, \mathrm{d}r = e^{-\mathrm{i}\pi/4} \int_0^{\infty} e^{-r^2} \, \mathrm{d}r = e^{-\mathrm{i}\pi/4} \sqrt{\pi}/2.$$

With this result, and that given on setting $-\mathrm{i}$ for i, we get

$$\int_{-\infty}^{\infty} e^{\pm\mathrm{i}\eta^2} \, \mathrm{d}\eta = 2 \int_0^{\infty} e^{\pm\mathrm{i}\eta^2} \, \mathrm{d}\eta = e^{\pm\mathrm{i}\pi/4} \sqrt{\pi}.$$

$$= \mathrm{Rl}\left\{\frac{2\pi}{\lambda|h''(t_0)|}\right\}^{\frac{1}{2}} e^{\mathrm{i}\lambda h(t_0)}\, e^{\pm \mathrm{i}\pi/4}\left\{1+O\left(\frac{1}{\sqrt{\lambda}}\right)\right\}.$$

Finally, from the last equation, the asymptotic form for (4.4) is thus

$$F(\lambda) = \int_a^b \cos \lambda h(t)\, \mathrm{d}t$$

$$\sim \left\{\frac{2\pi}{\lambda|h''(t_0)|}\right\}^{\frac{1}{2}} \cos\left(\lambda h(t_0) \pm \frac{\pi}{4}\right) + O\left(\frac{1}{\lambda}\right) \text{ as } \lambda \to \infty,$$

$$(4.11)$$

where $+\pi/4$ and $-\pi/4$ correspond respectively to $h''(t_0) > 0$ ($h(t_0)$ a minimum) and $h''(t_0) < 0$ ($h(t_0)$ a maximum).

If we now return to the original problem (4.1), we proceed in exactly the same way and, in this case, we take $g(t)$ into account by setting

$$g(t) = g(t_0) + (t-t_0)\, g'(t_0) + \ldots$$

in the vicinity of t_0 to obtain for (4.1) the analogous expression to (4.10), namely

$$f(\lambda) \sim \left\{\frac{2}{|h''(t_0)|}\right\}^{\frac{1}{2}} e^{\mathrm{i}\lambda h(t_0)} \int_{-s_1}^{s_1} e^{\pm \mathrm{i}\lambda s^2}\, g(t_0)\, \{1+O(s)\}\, \mathrm{d}s \text{ as } \lambda \to \infty,$$

and so, proceeding in the same way as we did above to obtain (4.11), we get

$$f(\lambda) = \int_a^b g(t)\, e^{\mathrm{i}\lambda h(t)}\, \mathrm{d}t \sim g(t_0)\left\{\frac{2\pi}{\lambda|h''(t_0)|}\right\}^{\frac{1}{2}} e^{\mathrm{i}(\lambda h(t_0) \pm \pi/4)} +$$

$$+ O\left(\frac{1}{\lambda}\right) \text{ as } \lambda \to \infty,$$

$$(4.12)$$

where again the positive and negative signs in the exponential correspond respectively to $h''(t_0) > 0$ and $h''(t_0) < 0$. Note that $|f(\lambda)| = O(1/\lambda^{\frac{1}{2}})$ as $\lambda \to \infty$ only as long as $g(t_0) \neq 0$.

We now return to the problem associated with the end points $t = a$ and $t = b$ mentioned above. At the same time, we consider the situation in which $h(t)$ has *no* turning point in $a \leqq t \leqq b$. In this latter case the only contributions to the integral (4.1) as $\lambda \to \infty$ are from the end points $t = a$ and $t = b$. We obtain the asymptotic expansions as $\lambda \to \infty$ immediately by integrating (4.1) by parts to give

$$\int_a^b g(t)\, e^{i\lambda h(t)}\, \mathrm{d}t = \int_a^b \frac{g(t)}{ih'(t)\lambda} \frac{\mathrm{d}}{\mathrm{d}t} \left\{ e^{i\lambda h(t)} \right\} \mathrm{d}t$$

$$= \frac{1}{\lambda} \left\{ \frac{g(b)}{ih'(b)}\, e^{i\lambda h(b)} - \frac{g(a)}{ih'(a)}\, e^{i\lambda h(a)} \right\} + O\left(\frac{1}{\lambda^2}\right) \text{ as } \lambda \to \infty. \tag{4.13}$$

In this case, $f(\lambda) = O(1/\lambda)$ as $\lambda \to \infty$. Thus if $h(t)$ also has a turning point in $a \leqq t \leqq b$, the contributions to the asymptotic expansion are $O(1/\lambda^{\frac{1}{2}})$ from the turning point, as in (4.12), and only $O(1/\lambda)$, as in (4.13), from the end points, and so in the turning point case (4.12) is the correct leading term as $\lambda \to \infty$.

Note, from (4.13), that the *relative* magnitudes of $h(a)$ and $h(b)$ are *not* of importance in the method of stationary phase, unlike the case in Laplace's method in §2.2. If it turns out that either end point is a genuine turning point, however, then that end point gives the dominant contribution and is given by $\frac{1}{2}$ of the right-hand side of (4.12) since the integration asymptotically will be from 0 to ∞ rather than $-\infty$ to ∞. For example, suppose $t = a$ is the only turning point; then $h'(a) = 0$ and

$$\int_a^b g(t)\, e^{i\lambda h(t)}\, \mathrm{d}t \sim g(a) \left\{ \frac{\pi}{2\lambda |h''(a)|} \right\}^{\frac{1}{2}} e^{i(\lambda h(a) \pm \pi/4)} \text{ as } \lambda \to \infty, \tag{4.14}$$

where $+\pi/4$ corresponds to $h''(a) > 0$ and $-\pi/4$ to $h''(a) < 0$.

A final point to note is that self-cancellation of oscillations, as is the case here, is a weaker decay mechanism than the exponential decay in §§2.1, 2.2, and 3.1, where contributions from other regions in the domain of integration were all *exponentially* smaller than the dominant term whereas here they are only *algebraically* smaller, as shown by comparing (4.13) with (4.12).

The procedure in the case where $h''(t_0) = 0$ or any number of derivatives are zero, is similar to that above, with appropriate modifications.

As an example, consider the asymptotic evaluation of the Bessel function of the first kind of integral order n, denoted by $J_n(x)$, as $x \to \infty$ with x real. This function can be defined (see, for example, Watson (1952)) by

$$J_n(x) = \frac{1}{\pi} \int_0^\pi \cos{(nt - x \sin t)}\, dt$$

$$= \frac{1}{\pi} \, \text{Rl} \int_0^\pi e^{int}\, e^{-ix\,\sin t}\, dt. \tag{4.15}$$

On comparison with (4.1), $h(t) = -\sin t$, $h'(t) = -\cos t$, and $h''(t) = \sin t$, giving $t_0 = \pi/2$ where $h'(t_0) = 0$, and hence $h(t_0) = -1$ and $h''(t_0) = 1 > 0$. Equation (4.12) is immediately applicable with $g(t_0) = e^{int_0} = e^{in\pi/2}$ and $\lambda = x$ to give

$$\frac{1}{\pi} \int_0^\pi e^{int}\, e^{-ix\,\sin t}\, dt \sim \frac{1}{\pi}\left(\frac{2\pi}{x}\right)^{\frac{1}{2}} e^{in\pi/2}\, e^{-ix}\, e^{i\pi/4},$$

and so, for (4.15),

$$J_n(x) \sim \left(\frac{2}{x\pi}\right)^{\frac{1}{2}} \cos\left(x - n\frac{\pi}{2} - \frac{\pi}{4}\right), \text{ as } x \to \infty, \tag{4.16}$$

which agrees with (3.71) if s there is an integer.

In the next section we briefly discuss linear dispersive wave motion as another illustrative example of the use of the method of stationary phase.

Exercises

1. Find the asymptotic expansion as $\lambda \to \infty$ of the following real integrals using the method of stationary phase:

   (i) $\displaystyle\int_0^1 \cos{\lambda(t^3 - t)}\, dt;$ (ii) $\displaystyle\int_{-\pi/2}^{\pi/2} \cos{(nt - \lambda \cos t)}\, dt$, n an integer;

   (iii) $\displaystyle\int_{-\infty}^{\infty} e^{i\lambda t^2}\, (1 + t^2)^{-1}\, dt.$

2. Fresnel's integrals are defined by

$$C(x) + iS(x) = \int_0^x e^{it^2}\, dt.$$

If

$$C(x) = \tfrac{1}{2}\left(\frac{\pi}{2}\right)^{\frac{1}{2}} - P(x)\cos x^2 + Q(x)\sin x^2,$$

$$S(x) = \tfrac{1}{2}\left(\frac{\pi}{2}\right)^{\frac{1}{2}} - P(x)\sin x^2 - Q(x)\cos x^2,$$

show that

$$P(x) \sim \frac{1}{4x^3}, \quad Q(x) \sim \frac{1}{2x} \quad \text{as} \quad x \to \infty.$$

3. (i) If $h(t)$ has a single stationary point at t_0, where $a < t_0 < b$, and $h'(t_0) = 0 = h''(t_0)$, $h'''(t_0) > 0$, show that

$$\int_a^b e^{i\lambda h(t)} \, dt \sim \Gamma\left(\frac{4}{3}\right) \left\{\frac{48}{\lambda h'''(t_0)}\right\}^{\frac{1}{3}} e^{i\lambda h(t_0)} \cos \pi/6 \text{ as } \lambda \to \infty;$$

(ii) Show that

$$\int_0^\pi e^{i\lambda(t - \sin t)} \, dt \sim \Gamma\left(\frac{4}{3}\right) \left(\frac{6}{\lambda}\right)^{\frac{1}{3}} e^{i\pi/6} \text{ as } \lambda \to \infty.$$

4. Find the asymptotic expansion as $\lambda \to \infty$ of

$$\int_0^1 e^{i\lambda t^s} \, dt, \, s \text{ real and } s > 1.$$

4.2. Linear dispersive wave motion and the method of stationary phase

Wave-like phenomena appear in an incredibly large number of different physical situations. Examples include surface waves on water, tides, certain recurring epidemics, electromagnetic waves, field oscillations in semiconductors, acoustic waves, shock wave propagation, car number-density waves in some traffic-flow models, atmosphere oscillations, vibrations in rods and so on. The spectrum of complexity of the governing equations also covers a wide range. However, in a large number of real situations the governing equation, or adequate approximate form of it, is not only linear but is the simplest either first- or second-order wave equation in two variables, x in space and t in time, namely

$$\phi_t + \phi_x = 0, \tag{4.17}$$

or

$$\phi_{xx} = \phi_{tt}, \tag{4.18}$$

where $\phi(x, t)$ denotes the disturbance. The subsequent discussion and philosophy do not of course, apply only to these equations. Specific examples are given at the end of this section which belong to neither of (4.17) nor (4.18).

An elementary harmonic wave train solution

$$\phi = a \cos (kx - wt) \qquad (4.19)$$

is found by substituting (4.19) into the equations (4.17) and (4.18) where a is the *amplitude*, $kx - wt$ is the *phase*, k is the *wave number* equal to $2\pi/l$, where l is the wave length, and w is the *frequency*, that is, $2\pi/w$ is the period of the wave. The form (4.19) is a solution of (4.18) if

$$w^2 = k^2, \qquad (4.20)$$

that is

$$w = k, \quad w = -k. \qquad (4.21)$$

With $w = k$, (4.19) is also a solution of (4.17). If x and t vary in such a way that $kx - wt = constant$, then on such a line in $x - t$ space ϕ is constant. The velocity defined by x/t is called the speed of propagation c of the disturbance ϕ since, if $x/t = w/k$, then for an observer moving with this speed c, $kx - wt = constant$ and hence, from (4.19), ϕ is constant. From (4.21) this means that $c = +1$ or $c = -1$, which represent, for (4.19), waves travelling in the positive and negative directions respectively. (This, of course, is a roundabout way to show that disturbances ϕ for (4.17) and (4.18) move with the characteristic speeds.) One point to note at this stage is that the speed of propagation of the wave (4.19), in which $w^2 = k^2$, is *independent* of the wave length k since $w \propto k$ and $c = w/k$.

More general elementary wave-like forms are

$$\phi = a(k) \cos (kx - w(k)t), \qquad (4.22)$$

where now the amplitude $a(k)$ and frequency $w(k)$ may be functions of k in general. The relationship between w and k—(4.20) is an example—is called the *dispersion* relation. There, with $w = \pm k$, the wave speed c is constant and so the wave disturbance is propagated *without* distortion in such a case. In the more general case (4.22) in which $w = w(k)$, the wave speed $c = w(k)/k$ is *not* constant with respect to k and so waves of different wave number move at different speeds which depend on the wave number. This type of situation is called *dispersive*. In such a system, if we are given an initial disturbance, the various waves making up the spectrum, as in a Fourier analysis, are continuously interacting with each other, since they move with different speeds. The wave train is thus a continuously changing form. For linear dispersive wave systems there is a reasonably complete theory†. Here we give a brief introduction to the basic ideas which illustrate an important use of the method of stationary phase.

With the fundamental form (4.22), we can get a more general form by assuming a Fourier integral form like

$$\phi(x, t) = \int_0^\infty a(k) \cos (kx - wt) \, dk, \qquad (4.23)$$

where here $a(k)$ is determined by the initial conditions. The specific differential equation is not important at this stage as long as it is linear.

As an example, suppose

$$\phi(x, 0) = H(x + X) - H(x - X), \qquad (4.24)$$

where $H(x)$ represents the Heaviside function

$$H(x) = 0 \text{ if } x < 0, \left.\right\} \\ = 1 \text{ if } x > 0. \qquad (4.25)$$

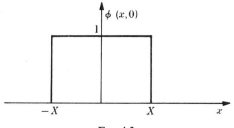

FIG. 4.3

Figure 4.3 illustrates $\phi(x, 0)$ in (4.24). On setting $t = 0$ in (4.23) and using Fourier's inversion theorem together with (4.24),

$$\phi(x, 0) = \frac{1}{\pi} \int_0^\infty \int_{-\infty}^\infty \phi(\xi, 0) \cos k\xi \cos kx \, d\xi \, dk$$

† See, for example, Jeffreys, H. and B. S. (1956). *Methods of mathematical physics* (3rd ed.). Cambridge University Press. Lighthill, M. J. (1965). *J. Inst. Math. Appl.* **1**, 1–28.

$$= \frac{1}{\pi} \int_0^\infty \int_{-x}^{x} \cos k\xi \cos kx \, d\xi \, dk$$

$$= \frac{2}{\pi} \int_0^\infty \frac{1}{k} \sin kX \cos kx \, dk, \tag{4.26}$$

which on comparison with (4.23) at $t = 0$ gives

$$a(k) = \frac{2}{\pi} \frac{\sin kX}{k} \tag{4.27}$$

We consider here those cases where $w(k)$ is a *real* function for real k. If w had an imaginary part, then we would also have dissipation which would cause decay (or growth) of the disturbance. To see this, if we write

$$\cos (kx - wt) = \text{Rl } e^{i(kx - wt)}$$

and $w = w_r + iw_i$, then e^{-iwt} involves $e^{w_i t}$ where w_i is real. If $w_i < 0$, the solution decays exponentially. If $w_i > 0$ the solution grows exponentially, which frequently indicates an instability in the problem.

To return to (4.23), $\phi(x, t)$ there with, for example, $a(k)$ in (4.27), is an explicit solution with (4.24) as the initial condition. In general it is diffcult to determine the main features of the solution. One thing we would like to know is what the wave train looks like after a long time, and whether there is an ultimate, or asymptotic in time, quasi-steady state. To this end, we must look for the asymptotic approximation of $\phi(x, t)$ for large t and large x. The reason for requiring large x as well as large t, is simply due to the fact that after a large time the waves have travelled a long distance. We shall now use the method of stationary phase to show that an *arbitrary initial disturbance*, of which (4.24) is an example, *disperses into a slowly varying wave train*.

Consider (4.23) when x and t are large but $x/t = O(1)$. If we now write $\phi(x, t)$ in (4.23) in the form

$$\phi(x, t) = \int_0^\infty a(k) \cos t \left(k \frac{x}{t} - w \right) dk, \tag{4.28}$$

we have the equivalent of the real part of (4.1) with t in place of λ, since $t \to \infty$, k in place of t, and $h(k) = kx/t - w(k)$. Since $h(k)$ is real, a turning point is given by a solution k_0 of

$$h'(k_0) = \frac{x}{t} - w'(k_0) = 0. \tag{4.29}$$

From (4.29), k_0 is given as a function of x/t. If this k_0 is such that $h''(k_0) = -w''(k_0) \neq 0$, we have, on comparison of (4.28) with (4.1) and (4.12),

$$\left.\begin{aligned} \phi(x, t) &\sim a(k_0) \left\{ \frac{2\pi}{t|w''(k_0)|} \right\}^{\frac{1}{2}} \cos\left(k_0 x - w(k_0)t \pm \frac{\pi}{4} \right) \text{ as } t \to \infty, \\ w'(k_0) &= x/t, \end{aligned}\right\} \tag{4.30}$$

where $+\pi/4$ corresponds in this case to $w''(k_0) < 0$ and $-\pi/4$ to the case $w''(k_0) > 0$: this is because of the way we defined $h(k)$ in terms of $w(k)$. If there are several positive k_0 solutions of the second equation of (4.30), there are the same number of contributions to $\phi(x, t)$ in the first equation of (4.30): they are all of $O(t^{-\frac{1}{2}})$.

The asymptotic result (4.30) illustrates some important, interesting phenomena about dispersive waves in general. The form (4.30) does represent an oscillatory wave form but, unlike the simple form (4.19), it is not uniform, since k_0 in (4.30) is a function of x/t. Since (4.30) holds for large x and large t, if we take the second equation of (4.30) and differentiate it with respect to x, we find

$$\frac{1}{k_0}\frac{dk_0}{dx} = \frac{w'(k_0)}{k_0 w''(k_0)}\frac{1}{x} = O\left(\frac{1}{x}\right) \ll 1, \tag{4.31}$$

which shows that the change in k_0 over a few wave lengths, that is a distance small compared with x, is *small* when x is large. With $x/t = O(1)$ in this analysis, (4.31) also shows that the change in k_0 over a few periods is also *small* when t is large. The result of this asymptotic analysis with a similar expression in t, thus shows, with (4.30) that locally in space and time the wave-like solution (4.30) is essentially an elementary solution like (4.19), with the a, k, and w given as *slowly* varying functions of x and t *over distances and times of order of a wave length and period*. The two scales, $1/k_0$ in length and $1/w(k_0)$ in time, are the order of the distances and times over which a, k, and w are essentially constant. One point to note in passing is that the *amplitude* of this slowly varying wave (4.30) *decays* like $O(t^{-\frac{1}{2}})$ for large times.

The velocity $w'(k_0)$ of (4.30) plays an important role in wave motion. It is the *group velocity*. From (4.30), it is the velocity, specifically x/t in the second equation of (4.30), that an observer would have to have in order to follow waves with a wave number in the vicinity of k_0: note the difference between it and the simple speed of propagation c.

This idea of a slowly varying wave train has been developed and exploited in a series of papers by Whitham[†] on nonlinear dispersive waves.

The specific dispersion relation relating w to k depends both on the differential equation governing the wave motion and the specific problem. For example, small-amplitude water waves in a fluid of depth H give rise to the dispersion relation[‡]

$$w^2(k) = gk \tanh kH, \tag{4.32}$$

where g is the acceleration due to gravity. On letting $H \to \infty$ in (4.32), we get the dispersion relation

$$w^2(k) = gk, \tag{4.33}$$

which corresponds to small-amplitude waves in a fluid of infinite depth. The group velocity here is $w'(k) = \frac{1}{2}(g/k)^{\frac{1}{2}}$.

As another example, suppose $\phi(x, t)$ satisfies

$$\phi_{tt} - \phi_{xx} + \phi = 0, \tag{4.34}$$

which represents an approximate linearized wave motion on a string, with a restoring force derived from a potential $\frac{1}{2}\phi^2$. The harmonic solution (4.19) is a solution of (4.34) if the dispersion relation

$$w(k) = (1 + k^2)^{\frac{1}{2}} \tag{4.35}$$

holds, as is seen immediately on substituting (4.19) into (4.34). In this case the group velocity is $w'(k) = k(1 + k^2)^{-\frac{1}{2}}$.

As a final example, consider the Korteweg–de Vries equation, which is the governing equation for long waves in shallow water, in the form

$$\phi_t + (1 + \varepsilon\phi)\,\phi_x + \mu\phi_{xxx} = 0, \tag{4.36}$$

---

† Whitham, G. B. (1970). *J. Fluid Mechs.* **44**, 373–395 (and the references given there).

‡ See, for example, Lamb, H. (1932). *Hydrodynamics*. Cambridge University Press (6th ed.) (also Dover Publications Inc., New York).

where ϕ is effectively proportional to the height of the free surface of the waves, ε and μ are small numbers, which reflect the fact that the depth of the water is small compared with the wave length of the long waves. This specific equation has been extensively studied, particularly in the last ten years. As it stands, we cannot apply the procedure we used for the equation (4.34), because (4.36) is a *non-linear* equation whereas (4.34) is linear. If we look for the wave pattern of the *linearized* form of (4.36), in which $\varepsilon\phi$ is neglected, we find that (4.19) is a solution with the dispersion relation

$$w(k) = k - \mu k^3, \tag{4.37}$$

with a group velocity $w'(k) = 1 - 3\mu k^2$. The asymptotic wave train obtained for the linearized form of (4.36), with $w(k)$ from (4.37), might be thought to be a reasonable first approximation to the exact solution of the nonlinear equation (4.36) since ε is small. In fact the asymptotic form of a wave train solution to the nonlinear equation is fundamentally different to that of the linearized equation. A partial reason is that although over a time scale $O(1)$ the small $\varepsilon\phi\phi_x$ term does have a small effect on the wave pattern, the effect of it over a *large* time is not small†.

Finally, as regards (4.36), the effect of the $\mu\phi_{xxx}$ term is *dispersive*, at least in the analysis we carried out on the linearized equation. If the term were $\mu\phi_{xx}$ instead, the problem is no longer dispersive, and solutions decay exponentially, as may be seen by considering the linearized version which is a parabolic differential equation like the heat equation.

---

† A similar large time effect may be seen by considering the simple equation $y'' + \varepsilon y' + y = 0$ for $y(t)$. Even if $0 < \varepsilon \ll 1$, no matter how small, we cannot neglect the $\varepsilon y'$ term over a long time, $O(1/\varepsilon)$ in fact. For example, the solution satisfying $y(0) = 0$, $y'(0) = 1$ is

$$y(t; \varepsilon) = \left\{1 - \left(\frac{\varepsilon}{2}\right)^2\right\}^{-\frac{1}{2}} e^{-\varepsilon t/2} \sin\left\{1 - \left(\frac{\varepsilon}{2}\right)^2\right\}^{\frac{1}{2}} t$$

$$\sim e^{-\varepsilon t/2} \{\sin t + O(\varepsilon^2)\}, \quad \varepsilon \ll 1.$$

When $\varepsilon \equiv 0$, $y(t; 0) = \sin t$ for *all* t. For $t = O(1)$ and $\varepsilon \ll 1$, $e^{-\varepsilon t/2} = 1 + O(\varepsilon)$ and so $y(t) = \sin t$ is a valid approximate solution to $O(1)$. However for $t \gg 1$, for $t \gg 1/\varepsilon$ in fact, $e^{-\varepsilon t/2} \ll 1$, and so for large t the solution is fundamentally different to that with $\varepsilon \equiv 0$: from the above solution $y(t; 0 < \varepsilon \ll 1) \to 0$ exponentially as $t \to \infty$.

5

Transform integrals

5.1. Transform integrals and their asymptotic evaluation

LINEAR boundary value problems are often solved by integral transform methods†. The final step in such a procedure is the transform inversion which involves, in general, the evaluation of integrals of the form

$$f(t) = \int_C F(z) \, K(z, t) \, dz, \tag{5.1}$$

where t is real, $K(z, t)$ is some given kernel, a function of two variables z and t, C is a given finite or infinite contour in the complex z-plane, and $F(z)$ is known, or at least its singularity properties are given. Here the function $F(z)$ is the integral transform of $f(t)$ and (5.1) defines its inverse.

The commonest transforms by far are the Fourier transform† and the Laplace transform†. In the Fourier transform case‡, the inversion equivalent to (5.1) is

$$f(t) = \frac{1}{\sqrt{(2\pi)}} \int_{-\infty}^{\infty} F(z) \, e^{itz} \, dz, \tag{5.2}$$

where t is real, *positive or negative*, and the specific contour is determined by the given problem. Here we consider the commonest

---

† See, for example, Carrier, G. F., Krook, M. and Pearson, C. E. (1966). *Functions of a complex variable*. McGraw Hill, New York. Sneddon, I. N. (1951). *Fourier transforms*. McGraw Hill, New York.

‡ The Fourier transform of $f(t)$ is here defined by $\dfrac{1}{\sqrt{(2\pi)}} \int_{-\infty}^{\infty} f(t) \, e^{-itz} \, dt$ and the Laplace transform by $\int_{0}^{\infty} f(t) \, e^{-zt} \, dt$

occurring contours, namely those which have end points at $\pm\infty$, by way of illustration. The Laplace transform‡ inversion equivalent to (5.1) is

$$f(t) = \frac{1}{2\pi i} \int\limits_{\alpha-i\infty}^{\alpha+i\infty} F(z)\, e^{tz}\, dz, \tag{5.3}$$

where t is real and *positive* and the contour here is from $\alpha-i\infty$ to $\alpha+i\infty$, where $\alpha > 0$ is such that all singularities of $F(z)$ lie to the left of the contour. Fourier sine and cosine transforms are simply special cases of (5.2) and their inversion formulae are

$$f(t) = \frac{1}{\sqrt{(2\pi)}} \int\limits_0^\infty F(z)\, \sin\, tz\, dz, \quad f(t) = \frac{1}{\sqrt{(2\pi)}} \int\limits_0^\infty F(z)\, \cos\, tz\, dz.$$

There are many more transforms, such as the Hilbert transform, Legendre transform, Hankel transform, and Mellin transform, for example, but these are less widely used. There are also finite transforms which involve finite contours.

In very many cases the exact inversion of the transform is not possible, and in many others where it can be done, the result is too complicated to be of much use. What can very often be obtained is the asymptotic evaluation of such transform integrals. The asymptotic approximation is frequently not only more useful than the exact answer (compare this with summing a divergent rather than a convergent asymptotic series) but is very often all that is required practically. The asymptotic result (4.30) of (4.28) in §4.2 is an example of this. Also, of course, such asymptotic evaluations are most useful in providing an accurate check for a numerical computation of the transforms. In general the above transform integrals are amenable to one or other of the asymptotic methods we have discussed: they may not necessarily be in the most convenient form for any given method, of course.

When transform methods are applied to partial differential equations, the possibility then exists of letting one or more of the independent variables tend to an asymptotic limit. In the simplest case of two independent variables r and t say, the function f in (5.1) to (5.3) is a function of r and t: in the integrand F is also a function of two variables.

In this section we consider the two most useful transforms, namely the Fourier and Laplace, from an asymptotic evaluation point of

‡ See the footnote on page 86.

view. We also make a few remarks on how the specific path of integration is chosen in a Fourier inversion integral. No new method for finding asymptotic approximations is discussed. The emphasis is on demonstrating the specific importance of the singularities on the asymptotic evaluation of the integrals (5.2) and (5.3), specifically as $t \to \pm\infty$. As might be expected from their form, the asymptotic method we ultimately use is Watson's lemma from §2.1.

The specific contour path in a Fourier transform inversion is implicitly contained in the statement of a well-posed problem amenable to transform methods. The contour must, of course, be chosen so that (5.2) converges. When $F(z)$ has branch points (which is the case more often than not) the position of the branch lines must be such as to ensure a unique answer with whatever *a priori* conditions are required on the solution. The actual determination of the contour and branch lines may be complicated, and is usually most easily handled by beginning with the probable or required asymptotic behaviour of the integral, and then finding which contour and branch lines give this behaviour. This is one of the main motivations for considering below the effect of different contours on the behaviour of transform integrals. Once the contour has been decided on, the results derived below also provide the asymptotic form for the transform integral in question.

Consider first an integral of the type (5.2), and let $F(z)$ have the appropriate behaviour for convergence of the integral and have singularities at $z = z_1$ and $z = Z_1$: these singularities may be either poles or branch points. To be specific, let us here take the contour C to go from $-\infty$ under z_1, and then over Z_1 to $+\infty$, as indicated in Fig. 5.1, and consider the consequences of this choice. If z_1 and Z_1 are branch points, let us suppose the branch lines are as shown, namely from z_1 to $Rl\,z_1 +i\infty$ and from Z_1 to $Rl\,Z_1 -i\infty$. These branch lines could be taken along other rays from z_1 and Z_1 as long as they do not cross C, of course. The specific path is crucially important in evaluating the integral, asymptotically, or otherwise.

The contour C may be deformed, by Cauchy's theorem, in any allowable way we wish, as long as the integral with the deformed contour exists, with whatever conditions were imposed on $F(z)$ at infinity in the original integral. Suppose $t < 0$ in (5.2). Then, with

$$z = x+iy, \quad z_1 = x_1+iy_1, \quad Z_1 = X_1+iY_1, \qquad (5.4)$$

if we tried to deform the contour C so that the ends lay at $x_1+i\infty$

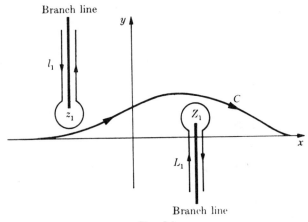

Branch line

FIG. 5.1

on either side of the branch line as denoted by l_1 in Fig. 5.1, $\left|e^{itz}\right| = \left|e^{-yt}\right| \to \infty$ as $y \to \infty$ since $t < 0$. Thus this deformation is *not* allowed since the resulting integral does not exist. We may, however, deform C to lie along L_1, as in Fig. 5.1, with end points at $X_1 - i\infty$ on either side of the branch line from Z_1, since, with $t < 0$, $\left|e^{itz}\right| = \left|e^{-yt}\right| \to 0$ as $y \to -\infty$. Suppose now $t > 0$. Then by a similar argument the contour C may be deformed to lie along l_1 in the upper-half plane in this case, but *not* along L_1 in the lower-half plane.

Consider first $t > 0$ and z_1 to be a pole, with r_1 the residue of $F(z)$ at z_1. The straightforward application of the theory of residues immediately gives, for (5.2) in this case, the exact expression

$$\frac{1}{\sqrt{(2\pi)}} \int_{-\infty}^{\infty} F(z)\, e^{itz}\, dz = (2\pi)^{\frac{1}{2}}\, i\, r_1\, e^{itz_1}, \tag{5.5}$$

where r_1 is the residue of $F(z)$ at the pole z_1, and $t > 0$.

Suppose now, with $t > 0$, that z_1 is a branch point of $F(z)$ such that, near z_1, we may write

$$F(z) = (z-z_1)^{\gamma_1} \sum_{n=0}^{\infty} a_n(z-z_1)^n, \quad a_0 \neq 0, \tag{5.6}$$

where γ_1 is real and, since the original integral exists, $\gamma_1 > -1$. Since $t > 0$ we may deform the integral to lie along l_1 as in Fig. 5.1. With (5.6), the integral round the circular arc at z_1 tends to zero in the limit as the radius tends to zero since $\gamma_1 > -1$, and so we have, for (5.2), on writing $z = z_1 + r e^{i\theta}$,

$$\frac{1}{\sqrt{(2\pi)}} \int_{-\infty}^{\infty} F(z) e^{itz} dz$$

$$= \frac{1}{\sqrt{(2\pi)}} \int_{l_1} F(z) e^{itz} dz$$

$$= \frac{1}{\sqrt{(2\pi)}} \int_{\infty}^{0} F(z_1 + r e^{-i3\pi/2}) e^{itz_1} e^{itre^{-i3\pi/2}} e^{-i3\pi/2} dr +$$

$$+ \frac{1}{\sqrt{(2\pi)}} \int_{0}^{\infty} F(z_1 + r e^{i\pi/2}) e^{itz_1} e^{itre^{i\pi/2}} e^{i\pi/2} dr$$

$$= \frac{i e^{itz_1}}{\sqrt{(2\pi)}} \int_{0}^{\infty} e^{-tr} \{F(z_1 + r e^{i\pi/2}) - F(z_1 + r e^{-i3\pi/2})\} dr. \qquad (5.7)$$

If we now consider the asymptotic form of (5.7) as $t \to \infty$, we use Watson's lemma from §2.1. This involves substituting (5.6) for the F's in the last integral, to give for (5.7)

$$\frac{1}{\sqrt{(2\pi)}} \int_{-\infty}^{\infty} F(z) e^{itz} dz \sim i \frac{e^{itz_1}}{\sqrt{(2\pi)}} \int_{0}^{\infty} e^{-tr} (r^{\gamma_1} e^{i\gamma_1\pi/2} \sum_{n=0}^{\infty} a_n r^n i^n -$$

$$- r^{\gamma_1} e^{-i\gamma_1 3\pi/2} \sum_{n=0}^{\infty} a_n r^n i^n) dr. \qquad (5.8)$$

Since, from §2.1,

$$\int_{0}^{\infty} e^{-tr} r^{\gamma_1+n} dr = t^{-(\gamma_1+n+1)} \Gamma(\gamma_1+n+1),$$

(5.8) finally becomes

$$\frac{1}{\sqrt{(2\pi)}} \int_{-\infty}^{\infty} F(z) e^{itz} dz \sim -\left(\frac{2}{\pi}\right)^{\frac{1}{2}} \frac{e^{i(tz_1-\gamma_1\pi/2)}}{t^{\gamma_1+1}} \sin \gamma_1\pi$$

$$\times \sum_{n=0}^{\infty} a_n \frac{i^n}{t^n} \Gamma(\gamma_1+n+1) \text{ as } t \to \infty,$$

$$(5.9)$$

where the γ_1 and a_n $(n = 0, 1, 2, \ldots)$ are defined by (5.6). The dominant term in the asymptotic expansion (5.9) is, for reference,

$$\frac{1}{\sqrt{(2\pi)}} \int_{-\infty}^{\infty} F(z) e^{itz} dz \sim -\left(\frac{2}{\pi}\right)^{\frac{1}{2}} a_0 \frac{e^{i(tz_1-\gamma_1\pi/2)}}{t^{\gamma_1+1}}$$

$$\times \Gamma(\gamma_1+1) \sin \gamma_1\pi \text{ as } t \to \infty. \qquad (5.10)$$

If $F(z)$ has several singularities at $z = z_i$, $i = 1, \ldots m$, say, above the contour C, the asymptotic evaluation involves the sum of m terms with contributions like (5.5) for a pole singularity and like (5.9) if the singularity in question is a branch point like z_1 in (5.6). From an asymptotic point of view as $t \to \infty$ such a sum of the various contributions from every singularity is really unnecessary since all that is required is the sum over those z_i with the *same smallest imaginary part* Im z_i, $i = 1, \ldots n$, because in (5.5) and (5.10) the exponential part is e^{-ty_1}, where $y_1 = \text{Im } z_1$.

Suppose now $t < 0$. The contour C must now be deformed to lie along the contour L_1 in Fig. 5.1, and in a similar manner to the above we must consider the asymptotic expansion resulting from the singularity at $z = Z_1$. When Z_1 is a pole of $F(z)$ with residue R_1, we have, in place of (5.5),

$$\frac{1}{\sqrt{(2\pi)}} \int_{-\infty}^{\infty} F(z) \, e^{itz} \, dz = -(2\pi)^{\frac{1}{2}} \, i \, R_1 \, e^{itZ_1}, \quad t < 0. \quad (5.11)$$

If Z_1 is a branch point near which

$$F(z) = (z - Z_1)^{\Gamma_1} \sum_{n=0}^{\infty} A_n (z - Z_1)^n, \quad (5.12)$$

then, analogous to (5.9), we have (exercise 1)

$$\frac{1}{\sqrt{(2\pi)}} \int_{-\infty}^{\infty} F(z) \, e^{itz} \, dz \sim -\left(\frac{2}{\pi}\right)^{\frac{1}{2}} \frac{e^{i(tZ_1 + \Gamma_1 \pi/2)}}{(-t)^{\Gamma_1 + 1}} \sin \Gamma_1 \pi$$

$$\times \sum_{n=0}^{\infty} A_n \frac{(-i)^n}{(-t)^n} \Gamma(\Gamma_1 + n + 1) \text{ as } t \to -\infty,$$

$$(5.13)$$

and the dominant term is, for reference,

$$\frac{1}{\sqrt{(2\pi)}} \int_{-\infty}^{\infty} F(z) \, e^{itz} \, dz \sim -\left(\frac{2}{\pi}\right)^{\frac{1}{2}} A_0 \frac{e^{i(tZ_1 + \Gamma_1 \pi/2)}}{(-t)^{\Gamma_1 + 1}}$$

$$\times \Gamma(\Gamma_1 + 1) \sin \Gamma_1 \pi \text{ as } t \to -\infty,$$

$$(5.14)$$

with A_0 and Γ_1 defined by (5.12). As before, if there are several relevant singularities Z_i, $i = 1, 2, \ldots M$, we simply sum the appropriate contributions from each according to the type of singularity. In this case the dominant asymptotic term is associated with the singularity Z_i which has the *largest imaginary part*.

Another point to note from these results is that *if a singularity z_i or Z_i lies on the imaginary axis, there is no oscillatory character from its asymptotic contribution* since e^{itz_i} if $t > 0$, is then simply the exponential e^{-ty_i} and $e^{itZ_i} = e^{-tY_i}$, if $t < 0$. It should be mentioned here, perhaps, that if there are no singularities below (above) the contour C and $t < 0$ ($t > 0$) the integral (5.2) is clearly identically zero. The Laplace transform (5.3) is only of interest, of course, for $t > 0$, by the same reasoning.

At this stage a few important points should be recalled about inversion paths in general. In any well-posed problem amenable to transform techniques, we must be able to specify the detailed path of the inversion contour. From the above results, it is clear that the path of the contour is absolutely crucial in evaluating the inversion integral. For example, if the contour C had passed below, rather than above, Z_1 in Fig. 5.1, with the branch line drawn upwards (so as not to cross C), the results (5.9) to (5.14) would have been completely different. The singularity at Z_1 would have contributed to the asymptotic expansion when $t > 0$ and *not* when $t < 0$. A fundamental principle is that if $F(z)$ is seemingly multi-valued, that is, if it has branch points, the inversion contour must be such that $F(z)$ is single-valued and uniquely given on the contour.

As an illustration of the above results and an amplification of these important points about transform solutions and their inversion paths, consider the asymptotic expansion as $|t| \to \infty$ of the integral

$$f(t) = \frac{1}{\sqrt{(2\pi)}} \int_{-\infty}^{\infty} e^{itz} (z-1)^{-1} (z^2 + 2i)^{-\frac{1}{2}} \, dz. \tag{5.15}$$

Here

$$F(z) = (z-1)^{-1} (z^2 + 2i)^{-\frac{1}{2}}, \tag{5.16}$$

which has three singularities, a simple pole at $z = z_1 = 1$ and two branch points, the solutions of $z^2 + 2i = 0$, at $z = z_2 = \sqrt{2} e^{-i\pi/4} = 1 - i$ and $z = z_3 = \sqrt{2} e^{i3\pi/4} = -1 + i$. To be specific, let us suppose, by way of example, that the required transform solution $f(t)$, given by (5.15), is such that it must tend to zero as $t \to \infty$. We now show how this determines uniquely the specific path of integration in (5.15). Only one of the several paths from $-\infty$ to ∞, some of which are given in Fig. 5.2, is the correct one which gives the appropriate behaviour for $f(t)$ at infinity.

We first show that the branch lines from z_2 and z_3 must be such that one must be drawn upwards and one downwards, by considering the multi-valued part of $F(z)$ in (5.16), namely $(z^2+2i)^{-\frac{1}{2}}$. If the contour and branch lines are as in Fig. 5.2(a) and (b), for example,

$$\arg (z^2+2i)^{-\frac{1}{2}} = -\tfrac{1}{2}\{\arg (z-z_2)+\arg (z-z_3)\}$$
$$\rightarrow -\tfrac{1}{2}(-\pi+\pi) = 0 \text{ as } z \rightarrow -\infty,$$

and

$$\arg (z^2+2i)^{-\frac{1}{2}} \rightarrow -\tfrac{1}{2}(0+0) = 0 \text{ as } z \rightarrow \infty,$$

and so $F(z)$ is on the same branch as $z \rightarrow \pm\infty$. If the branches from

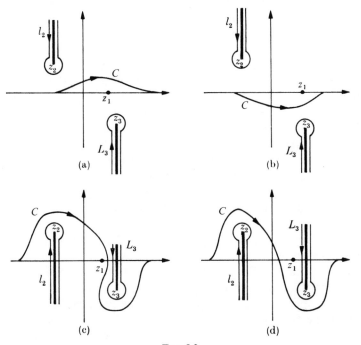

Fig. 5.2

z_2 and z_3 had both been drawn upwards, say, with the contour passing, necessarily of course, below z_2 and z_3

$$\arg (z^2+2i)^{-\frac{1}{2}} \rightarrow \pi \text{ as } z \rightarrow -\infty,$$
$$\rightarrow 0 \text{ as } z \rightarrow \infty,$$

and so $F(z)$ changes from one branch to another and is not uniquely given on the contour. A similar result is given if the contour passes above both z_2 and z_3. Thus the inversion contour must pass under one branch point and over the other. We are left, therefore, with the four basic possibilities shown in Fig. 5.2. To determine which of these contours is the appropriate one, we must evaluate for each of them the asymptotic behaviour of $f(t)$ in (5.15) as $t \to \infty$.

Consider first the contribution to $f(t)$ from the pole at $z = z_1 = 1$. If the contour passes under it the contribution to $f(t)$ in (5.15) involves, from (5.5), an oscillating term $O(e^{it})$ for all $t > 0$, and no contribution at all for $t < 0$. On the other hand, if the contour passes above $z = 1$ there is no contribution to $f(t)$ when $t > 0$ but there is the oscillatory term when $t < 0$. For the condition that $f(t) \to 0$ as $t \to \infty$, we must thus make the contour pass over the pole at $z = z_1 = 1$. We are now left with the two possibilities in Fig. 5.2 (a) and (c).

Consider first Fig. 5.2(a) and the contribution for $t > 0$ from the branch point at $z = z_2 = -1+i$. The expansion of $F(z)$ in (5.16) about the point z_2 in the form (5.6) gives $\gamma_1 = -\frac{1}{2}$, and so, from (5.10), there will be a dominant contribution $O(t^{-\frac{1}{2}} e^{itz_2})$ as $t \to \infty$. Since $z_2 = -1+i$, $|e^{itz_2}| = e^{-t} \to 0$ as $t \to \infty$: the contribution from the branch point at z_2 in Fig. 5.2(a) tends to zero as $t \to \infty$. If the contour is as in Fig. 5.2(c), the contribution to $f(t)$ for $t > 0$ is from the branch point at $z_3 = 1-i$, which gives a dominant contribution, from (5.10), $O(t^{-\frac{1}{2}} e^{itz_3}) = O(t^{-\frac{1}{2}} e^{it} e^t) \to \infty$ as $t \to \infty$. Thus the only contour which allows $f(t)$ to tend to zero as $t \to \infty$ is that in Fig. 5.2(a). The information we have used here, namely $f(t) \to 0$ as $t \to \infty$, is typical of that provided by a well-posed problem. With experience and practice, most of the above analysis can be dispensed with and the appropriate path decided on right away when the limiting value of $f(t)$ is given.

For completeness and illustration, we now evaluate the dominant term in the asymptotic expansion of $f(t)$ in (5.15) when the contour is as in Fig. 5.2(a). Since $F(z)$ is suitably behaved at infinity we may, as above, use Cauchy's theorem to deform the contour so that it lies along l_2 or L_3, as indicated in Fig. 5.2(a).

Consider first $t > 0$. There is only the one contribution from $z = z_2$. The required expansion of $F(z)$ about $z = z_2$ of the type (5.6) gives

$$
\begin{aligned}
a_0 &= (z_2 - 1)^{-1} (z_2 - z_3)^{-\frac{1}{2}} \\
&= (-1+i-1)^{-1} (-1+i-1+i)^{-\frac{1}{2}} \\
&= (5^{\frac{1}{2}} e^{i 5\pi/6})^{-1} (2^{\frac{3}{2}} e^{i 3\pi/4})^{-\frac{1}{2}} \\
&= 5^{-\frac{1}{2}} 2^{-\frac{3}{4}} e^{-i\pi 29/24}.
\end{aligned}
$$

Since, comparing $F(z)$ in (5.16) with (5.6), $\gamma_1 = -\frac{1}{2}$ and so $\Gamma(\gamma_1 + 1) = \Gamma(\frac{1}{2}) = \pi^{\frac{1}{2}}$, the dominant term in the expansion from (5.10), with a_0 as above, is

$$f(t) \sim -\left(\frac{2}{\pi}\right)^{\frac{1}{2}} 5^{-\frac{1}{2}} 2^{-\frac{3}{4}} e^{-i\pi 29/24} \pi^{\frac{1}{2}} t^{-\frac{1}{2}}$$

$$\times \quad e^{i(t(-1+i)+\pi/4)} \sin\left(-\frac{\pi}{2}\right).$$

Simplifying the last equation, we finally have, for the contour as in Fig. 5.2(a),

$$f(t) = \frac{1}{\sqrt{(2\pi)}} \int_{-\infty}^{\infty} e^{itz} (z-1)^{-1} (z^2 + 2i)^{-\frac{1}{2}} dz$$

$$\sim 5^{-\frac{1}{2}} 2^{-\frac{1}{4}} e^{-i\pi 23/24} t^{-\frac{1}{2}} e^{-t} e^{it} \text{ as } t \to \infty. \quad (5.17)$$

From (5.17), $f(t) \to 0$ exponentially as $t \to \infty$: it does so in an oscillatory manner.

Consider now $t < 0$. The pole at z_1 now contributes, and the residue R_1 of $F(z)$ at $z = 1$ is, from (5.16), $R_1 = (1 + 2i)^{-\frac{1}{2}} = 5^{-\frac{1}{4}} e^{-i\pi/6}$, and so its contribution to $f(t)$ is, from (5.11),

$$-(2\pi)^{\frac{1}{2}} i 5^{-\frac{1}{4}} e^{-i\pi/6} e^{it} = \pi^{\frac{1}{2}} 2^{\frac{1}{2}} 5^{-\frac{1}{4}} e^{i(t+4\pi/3)}. \quad (5.18)$$

The contribution from the branch point at $z = z_3 = 1 - i$ is obtained from (5.14). Here, comparing $F(z)$ in (5.16) near $z = z_3$ with (5.12),

$$\Gamma_1 = -\frac{1}{2}, \quad A_0 = (z_3 - 1)^{-1} (z_3 - z_2)^{-\frac{1}{2}}$$

$$= (1 - i - 1)^{-1} (1 - i + 1 - i)^{-\frac{1}{2}} = 2^{-\frac{3}{4}} e^{i\pi 5/8},$$

and so its contribution as $t \to -\infty$, from (5.14), is

$$-\left(\frac{2}{\pi}\right)^{\frac{1}{2}} 2^{-\frac{3}{4}} e^{i\pi 5/8} \pi^{\frac{1}{2}} (-t)^{-\frac{1}{2}} e^{i(t(1-i)-\pi/4)} \sin\left(-\frac{\pi}{2}\right)$$

$$= 2^{-\frac{1}{4}} e^{i(t+3\pi/2)} (-t)^{-\frac{1}{2}} e^{t}. \quad (5.19)$$

As $t \to -\infty$, (5.19) tends to zero exponentially, whereas (5.18) is bounded and oscillatory. Thus as $t \to -\infty$, the dominant term in the asymptotic expansion of (5.15), with the contour as in Fig. 5.2(a), is, using (5.18), finally

$$f(t) = \frac{1}{\sqrt{(2\pi)}} \int_{-\infty}^{\infty} e^{itz} (z-1)^{-1} (z^2+2i)^{-\frac{1}{2}} dz$$

$$\sim \pi^{\frac{1}{2}} 2^{\frac{1}{2}} 5^{-\frac{1}{4}} e^{i(t+\pi 4/3)} \text{ as } t \to -\infty, \tag{5.20}$$

which is bounded and purely oscillatory.

If we now consider the Laplace integral (5.3), we may proceed in a similar manner to that used for Fourier integrals. Here, however, the contour is specified exactly. Fourier and Laplace transforms are, of course, closely related. In keeping with the inversion forms (5.2) and (5.3), the Fourier transform of $f(t)$ is defined by

$$\frac{1}{\sqrt{(2\pi)}} \int_{-\infty}^{\infty} f(t) e^{-itz} dt$$

and the Laplace transform by $\int_{0}^{\infty} f(t) e^{-tz} dt$ (see the footnote on page 86). If we write iz for z in the Laplace transform of $f(t)$, it becomes $\sqrt{(2\pi)}$ times the Fourier transform of a function which is the same as $f(t)$ for $t > 0$ and which is zero for $t < 0$.

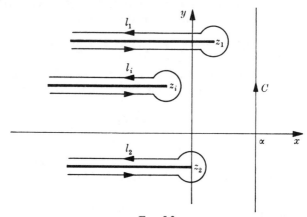

FIG. 5.3

Let z_i, $i = 1, 2, \ldots n$ be the singularities of $F(z)$ of (5.3) as indicated in Fig. 5.3. Here $\alpha > \text{Rl } z_i$ for all $i = 1, 2, \ldots n$ and $F(z)$ is suitably bounded at infinity so that we may use Cauchy's theorem to deform the path C so that it lies along the contours l_i, $i = 1, \ldots n$, where now the branch lines from those of the singularities which are branch points are drawn parallel to the negative real axis, from the branch point to infinity, as shown. The dominant asymptotic term now comes from the singularity with the *largest real part* since from each singularity z_i we have (compare with (5.5), (5.9), (5.11), and (5.13)), in the asymptotic evaluation, the exponential e^{tz_i}, and with z_1, say, as the singularity with the largest real part $e^{tz_i} = o(e^{tz_1})$ for all $i = 2, 3, \ldots n$.

Suppose for simplicity that z_1 is the only singularity of $F(z)$. If z_1 is a pole, we have for (5.3), by Cauchy's theorem, simply

$$\frac{1}{2\pi i} \int_{\alpha-i\infty}^{\alpha+i\infty} F(z) \, e^{tz} \, dz = r_1 \, e^{tz_1}, \quad t > 0, \tag{5.21}$$

where r_1 is the residue of $F(z)$ at $z = z_1$.

If z_1 is a branch point of $F(z)$ which has an expansion of the form (5.6), then proceeding as above (compare with (5.7) and (5.9)) we get for its contribution

$$\frac{1}{2\pi i} \int_{\alpha-i\infty}^{\alpha+i\infty} F(z) \, e^{tz} \, dz$$

$$= \frac{1}{2\pi i} e^{tz_1} \int_0^\infty F(z_1 + r \, e^{-i\pi}) \, e^{-tr} \, dr - \frac{1}{2\pi i} e^{tz_1} \int_0^\infty F(z_1 + r \, e^{i\pi}) \, e^{-tr} \, dr$$

$$\sim -\frac{1}{\pi} e^{tz_1} \sin \gamma_1 \pi \sum_{n=0}^\infty a_n (-1)^n \int_0^\infty e^{-tr} r^{\gamma_1 + n} \, dr$$

$$= -\frac{1}{\pi} \frac{e^{tz_1}}{t^{\gamma_1+1}} \sin \gamma_1 \pi \sum_{n=0}^\infty a_n (-1)^n t^{-n} \Gamma(\gamma_1 + n + 1) \text{ as } t \to \infty. \tag{5.22}$$

The dominant term of (5.3) is thus

$$\frac{1}{2\pi i} \int_{\alpha-i\infty}^{\alpha+i\infty} F(z) \, e^{tz} \, dz \sim -\frac{a_0}{\pi} \frac{e^{tz_1}}{t^{\gamma_1+1}} \Gamma(\gamma_1 + 1) \sin \gamma_1 \pi \text{ as } t \to \infty. \tag{5.23}$$

The situation when there are n singularities is clearly just a sum of appropriate terms like (5.21) for a pole, and (5.22) for a branch point, with an expansion for $F(z)$, near it, like (5.6).

The forms (5.21) and (5.23) immediately confirm the above statement that the singularity with the largest real part dominates the asymptotic expansion.

In the Laplace inversion situation a singularity which lies on the imaginary axis gives a purely oscillatory contribution, whereas if it lies on the real axis it gives a decaying or growing exponential contribution depending on whether it is negative or positive respectively.

Exercises

1. With $F(z)$ given by (5.12), show that the asymptotic evaluation of the Fourier inversion (5.2) in the case where $t \to -\infty$ is given by (5.13).

2. Obtain the dominant term in the asymptotic evaluation as $|t| \to \infty$ of the Fourier inverse transform

$$\frac{1}{\sqrt{(2\pi)}} \int_{-\infty}^{\infty} e^{itz} (z^2 - 1)^{-1} \, dz.$$

   when the contour passes (i) below $z = -1$ and above $z = +1$, and (ii) above $z = -1$ and below $z = +1$.

3. Find the appropriate inversion path for the following Fourier inversion if we require $f(t) \to 0$ as $t \to \infty$, and hence obtain the asymptotic expansion as $t \to \pm \infty$:

$$\frac{1}{\sqrt{(2\pi)}} \int_{-\infty}^{\infty} e^{itz} (z-1)^{-\frac{1}{2}} (z+1+i)^{-\frac{1}{2}} (z^2+1)^{-1} \, dz.$$

4. Obtain the dominant term in the asymptotic evaluation as $t \to \infty$ of the following Laplace transform inversions:

   (i) $\dfrac{1}{2\pi i} \displaystyle\int_{\alpha-i\infty}^{\alpha+i\infty} e^{tz} z^{-1} \tanh z \, dz$

   (ii) $\dfrac{1}{2\pi i} \displaystyle\int_{\alpha-i\infty}^{\alpha+i\infty} e^{tz} z^{-\frac{1}{2}} e^{-z^{\frac{1}{2}}} \, dz.$

5. Show that

$$\frac{1}{2\pi i} \int_{\alpha-i\infty}^{\alpha+i\infty} e^{tz} (z^2 + 2z + 2)^{-1} (z^{\frac{1}{2}} + 1)^{-1} \, dz \sim \frac{1}{4\pi^{\frac{1}{2}} t^{\frac{3}{2}}}$$

   as $t \to \infty$ when $|\arg z| < \pi$. (Note that with this argument range there is no pole from $(z^{\frac{1}{2}} + 1)^{-1}$).

6

Differential equations

6.1. Singularities and asymptotic methods of solution

THE methods in the previous sections provide asymptotic approximations for functions defined by integrals. Many of these functions are solutions of specific differential equations: the Bessel function is an example. In the case of functions defined as the solutions of differential equations which cannot be solved explicitly, these integral methods described above naturally cannot be used. (We group in the class of explicit solutions those given in the form of an integral.) In these circumstances we must resort to differential equation methods to find asymptotic approximations for the functions. The subject of asymptotic methods for solving differential equations is large. In this and the following sections we give a brief introduction to the subject. In the case of ordinary differential equations the book by Wasow (1965) discusses several aspects with detail and rigour. A large part of each of the books by Erdelyi (1956) and Jeffreys (1966) is devoted to ordinary differential equations. In the partial differential equation area the books by Van Dyke (1964) and Cole (1968)† are of importance in the particular area of asymptotic analysis called singular perturbation theory.

Here we shall discuss linear second-order homogeneous differential equations for $w(z)$ of the form

---

† Van Dyke, M. (1964). *Perturbation methods in fluid mechanics*. Academic Press, New York.
Cole, J. D. (1968). *Perturbation methods in applied mathematics*. Blaisdell Publishing Co., Waltham, Mass.

$$w''(z) + p\,w'(z) + q\,w(z) = 0 \qquad (6.1)\dagger$$

where z is in general complex and, with the possible exception of the point at infinity, p and q are analytic functions of z and any parameters of the problem. We wish to find the asymptotic expansion for $w(z)$ as $z \to \infty$, or as some parameter tends to infinity, by methods other than finding an exact explicit solution of (6.1) and *then* considering the asymptotic form of it. If the limiting point in question is not $z = \infty$, we make the appropriate transformation to make it so. In what follows it should be kept in mind that asymptotic expansions are usually only valid for some restricted z-domain.

A more convenient form of (6.1) for study is obtained on setting

$$w(z) = W(z)\,e^{-\frac{1}{2}\int p(z)\,dz}, \qquad (6.2)$$

which gives in place of (6.1)

$$W'' + (q - \tfrac{1}{2}p' - \tfrac{1}{4}p^2)\,W = 0. \qquad (6.3)$$

Thus, without loss of generality, we may consider at the outset

$$w'' + f(z)\,w = 0, \qquad (6.4)$$

where $f(z)$ is, for the moment, an analytic function of z, except possibly at infinity. $f(z)$ is related to p and q of (6.1) using (6.3). In (6.4) f may also depend on an independent parameter, λ say, in which case we write it as $f(z; \lambda)$ when we wish to emphasize the fact, as in (6.43) in §6.2. We discuss here two distinct types of asymptotic problems associated with such differential equations, namely those in which the independent variable $z \to \infty$ (this section) and those in which an independent parameter $\lambda \to \infty$ or $\lambda \to 0$ (§§6.2 and 6.3). This latter class illustrates one aspect of singular perturbation theory.

As a preliminary, we must discuss the singularities, considered here to be isolated, which $w(z)$ will have as a consequence of the behaviour of p and q as $z \to \infty$.

If $z = \infty$ is an ordinary point of (6.4), then as $z \to \infty$, $w(z)$ consists of two independent series expansions, in inverse powers of z, which are convergent for $|z| > R$ for some R. If the point at infinity is not an ordinary point, it is a *singularity* which is usually one of two kinds.

The point $z = \infty$ is a *regular singularity* of $w(z)$ if $f(z) = O(z^{-2})$ as $z \to \infty$: in terms of p and q, this means (from (6.3)) that $p = O(z^{-1})$

---

† Throughout this and the following sections we shall frequently just use the letter for the functions, for example, p for $p(z)$ and so on.

and $q = O(z^{-2})$ as $z \to \infty$. In fact, we require $z^2 f$, zp, $z^2 q$ to be analytic as $z \to \infty$. In this case the usual Frobenius procedure† of substituting

$$w(z) = z^\rho \sum_{n=0}^{\infty} c_n z^{-n}, \tag{6.5}$$

into (6.4) (or (6.1)) and equating like powers of z, determines ρ and c_n, $n = 0, 1, 2, \ldots$. The coefficient of the lowest power of z set equal to zero gives the indicial equation which determines two values for ρ and hence for each of the c_n. When the two values of the ρ's are not integers, the solutions in (6.5) are not analytic since $z = \infty$ is a branch point. If the roots of the indicial equation differ by an integer, then one of the two solutions for $w(z)$ may possess a $\log z$ term: if the two ρ's are equal then a $\log z$ term always appears. Again the solutions $w(z)$ are not analytic at infinity. The solutions obtained in this way are asymptotic solutions for $w(z)$ as $z \to \infty$. We do not discuss this type of singularity further. The reader is referred to any of the books† on ordinary differential equations.

The point at infinity is an *irregular singularity* of $w(z)$ if $f(z)$ is *not* $O(z^{-2})$ as $z \to \infty$. Examples are $f(z) = O(1)$, $f(z) = O(z)$, or $f(z) = O(z^{-1})$, and so on, as $z \to \infty$. Equations with irregular singularities are discussed in this chapter. In this section we describe in detail two practical methods of solution, the second of which is less standard, more basic but yet simpler, *and* has wider applications (see also §6.2).

From a pedagogical point of view, we first motivate the standard method by considering $f(z)$ in (6.4) to have an asymptotic form

$$f(z) \sim a_0 + \frac{a_1}{z^2} + \ldots = a_0 + O\left(\frac{1}{z^2}\right) \quad \text{as } z \to \infty, \tag{6.6}$$

where $a_0 \neq 0$. A term $O(z^{-1})$ is omitted here for simplicity since it is not necessary for motivating the general procedure below. Such a term is included in both of the general methods, where its effect, which is important, is pointed out. If (6.6) is substituted into (6.4), we might reasonably expect the asymptotic expansion for $w(z)$ in the

---

† See, for example, Burkill, J. C. (1956). *The theory of ordinary differential equations*. Oliver and Boyd, Edinburgh. Ince, E. L. (1965). *Ordinary differential equations*. Dover Publications Inc., New York.

form

$$w(z) \sim w_0(z) + w_1(z) + \ldots \tag{6.7}$$

to be obtained by solving successively for w_0, w_1, ..., where $w_n = o(w_{n-1})$. Equation (6.4), with (6.6) and (6.7), becomes

$$(w_0'' + w_1'' + \ldots) + \left(a_0 + \frac{a_2}{z^2} + \ldots\right)(w_0 + w_1 + \ldots) \sim 0. \tag{6.8}$$

The dominant term $w_0(z)$ then satisfies

$$w_0'' + a_0 \, w_0 = 0 \Rightarrow w_0(z) = A_0 \, e^{ia_0^{1/2}z} + B_0 \, e^{-ia_0^{1/2}z}, \tag{6.9}$$

which now gives, from (6.8),

$$w_1'' + a_0 \, w_1 = -\frac{a_2}{z^2} \, w_0 = O(z^{-2} \, e^{\pm ia_0^{1/2}z}). \tag{6.10}$$

Since, in our scheme (6.7), we require $w_1 = o(w_0)$, we need only the particular integral of (6.10) which is

$$w_1(z) = O(z^{-1} \, e^{\pm ia_0^{1/2}z}). \tag{6.11}$$

Thus (6.9) and (6.11) together give

$$w(z) = e^{ia_0^{1/2}z} \{A + O(z^{-1})\} + e^{-ia_0^{1/2}z} \{B + O(z^{-1})\}, \tag{6.12}$$

where A and B are constants.

While discussing and using these general methods below, we should remember that the domain of z in the asymptotic expansion used for $f(z)$ as $z \to \infty$ clearly puts limitations on the asymptotic solutions $w(z)$ as $z \to \infty$. In many cases $f(z)$ is analytic at infinity and its asymptotic expansion is simply a convergent series in inverse powers of z for large enough z. Of course $w(z)$ is *not* analytic at infinity: it has an irregular singularity there whenever $f(z)$ is not $O(z^{-2})$.

Keeping in mind the form (6.5) for a regular singularity together with the above heuristic approach, which resulted in (6.12), suggests that a more general procedure for obtaining asymptotic solutions for (6.4) with

$$f(z) \sim a_0 + \frac{a_1}{z} + \ldots \quad \text{as } z \to \infty \tag{6.13}$$

is to consider

$$\left. \begin{array}{l} w(z) = e^{\lambda z} \, z^\sigma \, g(z) \\[2mm] g(z) \sim \alpha_0 + \dfrac{\alpha_1}{z} + \ldots \end{array} \right\} \quad \text{as } z \to \infty, \tag{6.14}$$

where the coefficients a_n, $n = 0, 1, 2, \ldots$ are obtained from $f(z)$ in (6.13) and the coefficients α_n, $n = 0, 1, 2, \ldots$, λ, and σ in (6.14) have to be found. Substituting (6.13) and (6.14) into (6.4), we now obtain, by successive approximations two asymptotic solutions for $w(z)$ as $z \to \infty$. From (6.14),

$$w''(z) = e^{\lambda z}\, z^{\sigma} \left\{ \alpha_0\, \lambda^2 + \frac{1}{z}\, (\lambda^2\, \alpha_1 + 2\lambda\sigma\, \alpha_0) + \ldots \right\},$$

which, on substituting into (6.4), gives, on cancelling $e^{\lambda z}\, z^{\sigma}$ and collecting terms of like powers in z,

$$\left\{\alpha_0(\lambda^2 + a_0)\right\} + \frac{1}{z}\left\{\alpha_1(\lambda^2 + a_0) + \alpha_0(2\lambda\sigma + a_1)\right\} + \ldots = 0.$$

$$(6.15)$$

Successively applying the asymptotic limit $z \to \infty$ to (6.15), which is equivalent to setting the coefficient of each power of z to zero, we get two possible values for λ and hence two sets of values for σ and α_n, $n = 1, 2, \ldots$. If $a_0 \neq 0$, for example, (6.15) gives, on equating powers of $z^{-1}, z^{-2}, \ldots z^{-n}, \ldots$

$$\left. \begin{array}{l} \lambda^{(1)}, \lambda^{(2)} = \pm i a_0^{\frac{1}{2}}, \quad \sigma^{(1)}, \sigma^{(2)} = -\dfrac{a_1}{2\lambda} = \pm\dfrac{i a_1}{2 a_0^{\frac{1}{2}}}, \\[3mm] \alpha_1^{(1)}, \alpha_1^{(2)} = \mp\dfrac{i\alpha_0}{2 a_0^{\frac{1}{2}}}\left(\mp\dfrac{i a_1}{2 a_0^{\frac{1}{2}}} - \dfrac{a_1^2}{4 a_0} + a_2\right), \\[3mm] \alpha_{n+1} = \dfrac{1}{2\lambda(n+1)}\left[\left\{n(n+1) - (2n+1)\,\sigma + \sigma^2 + a_2\right\}\alpha_n + \right. \\[3mm] \left. + a_3\,\alpha_{n-1} + \ldots + a_{n+2}\,\alpha_0\right], n = 0, 1, 2, \ldots, \end{array} \right\} \quad (6.16)$$

with the convention that symbols with negative subscripts, α_{-1} for example, are zero.

In (6.16) α_0 is arbitrary, as it must be in finding a general solution to the homogeneous equation (6.4). Note that in (6.16) there are two λ's which determine the two corresponding σ's, and two corresponding values for each α_n and hence two asymptotic solutions for $w(z)$ as $z \to \infty$, namely

$$\left. \begin{array}{l} w_1(z) \sim A\, e^{\lambda^{(1)} z}\, z^{\sigma^{(1)}}\left(1 + \dfrac{\alpha_1^{(1)}}{\alpha_0}\dfrac{1}{z} + \ldots + \dfrac{\alpha_n^{(1)}}{\alpha_0}\dfrac{1}{z^n} + \ldots\right), \\[4mm] w_2(z) \sim B\, e^{\lambda^{(2)} z}\, z^{\sigma^{(2)}}\left(1 + \dfrac{\alpha_1^{(2)}}{\alpha_0}\dfrac{1}{z} + \ldots + \dfrac{\alpha_n^{(2)}}{\alpha_0}\dfrac{1}{z^n} + \ldots\right), \end{array} \right\} (6.17)$$

where A, B are arbitrary constants and $\lambda^{(1)}$, $\lambda^{(2)}$, $\sigma^{(1)}$, $\sigma^{(2)}$, $\alpha_n^{(1)}$, $\alpha_n^{(2)}$, $n = 1, 2, \ldots$, are given by (6.16), the last of which gives $\alpha_n^{(1)}$, for example, when $\lambda^{(1)}$, $\sigma^{(1)}$, $\alpha_{n-1}^{(1)}$, and so on are used. The connection between (6.17) and the solutions obtained in (6.12) is now clear from (6.16). When $f(z) \sim a_0 + O(z^{-2})$, as in (6.6), that is $a_1 = 0$ in (6.13), $\sigma = 0$ from (6.16) and (6.17) then gives (6.12).

With (6.17) we can see that if we terminate either series at $n = N$, say, we have

$$w(z) = C\, e^{\lambda z}\, z^\sigma \left(1 + \frac{\alpha_1}{\alpha_0}\frac{1}{z} + \ldots + \frac{\alpha_N}{\alpha_0}\frac{1}{z^N}\right) + O(e^{\lambda z}\, z^{\sigma - N - 1}),$$

$$(6.18)$$

where C is an arbitrary constant. Substituting (6.18) into (6.4) gives

$$w'' + f(z)\, w = O(e^{\lambda z}\, z^{\sigma - N - 3}), \qquad (6.19)$$

which shows that the asymptotic error in (6.4), on taking (6.18) to the order given as the solution, is $O(e^{\lambda z}\, z^{\sigma - N - 3})$ as compared with $O(e^{\lambda z}\, z^{\sigma - N - 1})$.

Suppose now that $f(z)$ is such that $a_0 = 0$, $a_1 \neq 0$ in (6.13) (solution (6.12) has $a_0 \neq 0$, $a_1 = 0$). In this case we may write (6.4) as

$$\left.\begin{array}{l} w'' + \dfrac{1}{z}\, F(z)\, w = 0, \\[2em] F(z) \sim a_1 + \dfrac{a_2}{z} + \ldots \text{ as } z \to \infty. \end{array}\right\} \qquad (6.20)$$

If we put $z = Z^2$, $w = Z^{\frac{1}{2}}\, W(Z)$, (6.20) becomes

$$\left.\begin{array}{l} W''(z) + G(Z)\, W(Z) = 0, \\[1.5em] G(Z) = 4F(Z^2) - \dfrac{3}{4Z^2} \sim 4a_1 + O\left(\dfrac{1}{Z^2}\right), \end{array}\right\}$$

which is the same as the above (6.6), with $f(z) \sim a_0 + O(1/z^2)$ and hence (6.18) with $\sigma = 0$ (or (6.12)) applies. Thus

$$W(Z) = C\, e^{\pm i(4a_1)^{\frac{1}{2}} Z}\left\{1 + O\left(\frac{1}{Z}\right)\right\}$$

and so, for (6.20)

$$w(z) = C\, z^{\frac{1}{4}}\, e^{\pm 2i(a_1)^{\frac{1}{2}} z^{\frac{1}{2}}}\left\{1 + O\left(\frac{1}{z^{\frac{1}{2}}}\right)\right\}. \qquad (6.21)$$

The case when $a_0 = 0 = a_1$, $a_2 \neq 0$, is of course the regular singularity case covered by (6.5).

As a straightforward example of the procedure which uses (6.14), we consider the Bessel equation, with real argument x,

$$\phi'' + \frac{1}{x}\phi' + \left(1 - \frac{v^2}{x^2}\right)\phi = 0, \tag{6.22}$$

where v is a real constant. The transformation (6.2), from ϕ to w say, with $p = 1/x$, gives

$$\phi(x) = w(x)\,e^{-\frac{1}{2}\int x^{-1}dx} = x^{-\frac{1}{2}}\,w(x),$$

and (6.22) becomes

$$w''(x) + \left(1 - \frac{v^2 - \frac{1}{4}}{x^2}\right)w(x) = 0, \tag{6.23}$$

and so the equivalent of f in (6.4) is

$$f(x) = 1 - \frac{v^2 - \frac{1}{4}}{x^2} = O(1) \text{ as } x \to \infty.$$

Thus, comparing this with (6.13),

$$a_0 = 1, \quad a_1 = 0, \quad a_2 = -(v^2 - \tfrac{1}{4}), \quad a_{n \geq 3} = 0. \tag{6.24}$$

Substitution of (6.14), with (6.24), into (6.23) gives for (6.15)

$$\{\alpha_0(\lambda^2 + 1)\} + \frac{1}{x}\{\alpha_1(\lambda^2 + 1) + \alpha_0\,2\lambda\sigma\} + \ldots = 0,$$

which on equating to zero the coefficients of x^{-n}, $n = 0, 1, \ldots$, we get for (6.16)

$$\left. \begin{array}{l} \lambda^{(1)}, \lambda^{(2)} = \pm i, \ \sigma = 0, \ \alpha_1^{(1)}, \alpha_1^{(2)} = \pm\dfrac{i\alpha_0}{2}(v^2 - \tfrac{1}{4}), \\[3mm] \alpha_{n+1}^{(1)}, \alpha_{n+1}^{(2)} = \mp\dfrac{i}{2(n+1)}\{n(n+1) - (v^2 - \tfrac{1}{4})\}\,\alpha_n^{(1)}, \alpha_n^{(2)}, \ n = 0, 1, \ldots \end{array} \right\} \tag{6.25}$$

The two independent solutions for $w(x)$ are, using (6.25), now given by

$$\left. \begin{array}{l} w_1(x) = A\,e^{ix}\left\{1 + \dfrac{i(v^2 - \frac{1}{4})}{2x} + O\left(\dfrac{1}{x^2}\right)\right\}, \\[4mm] w_2(x) = B\,e^{-ix}\left\{1 - \dfrac{i(v^2 - \frac{1}{4})}{2x} + O\left(\dfrac{1}{x^2}\right)\right\}. \end{array} \right\} \tag{6.26}$$

The two solutions for $\phi(x)$ in (6.22) are given by $\phi(x) = x^{-\frac{1}{2}}\,w(x)$, with $w(x)$ each of the two solutions in (6.26). If we choose the specific values of A and B in (6.26) as

$$A = \frac{1}{\sqrt{(2\pi)}} e^{-i(\pi/2)\,(\nu+\frac{1}{2})}, \quad B = \frac{1}{\sqrt{(2\pi)}} e^{i(\pi/2)\,(\nu+\frac{1}{2})}, \quad (6.27)$$

we have as a solution of (6.22)

$$\phi(x) = x^{-\frac{1}{2}} \{A\, w_1(x) + B\, w_2(x)\}$$

$$\sim \frac{1}{(2\pi x)^{\frac{1}{2}}} \left[e^{i\{x-(\pi/2)\,(\nu+\frac{1}{2})\}} + e^{-i\{x-(\pi/2)\,(\nu+\frac{1}{2})\}} \right] + \cdots$$

$$= \left(\frac{2}{\pi x}\right)^{\frac{1}{2}} \cos\left(x - \frac{\pi}{2}\,\nu - \frac{\pi}{4}\right) + O(x^{-\frac{3}{2}}), \quad (6.28)$$

which, with λ for x and s for ν, is the asymptotic expansion for $J_s(\lambda)$ as $\lambda \to \infty$ obtained in §3.2 and given by (3.71).

An alternative procedure, which has important extensions (two of which we develop in §6.2), for finding asymptotic solutions of (6.4) with $f(z)$ as in (6.13), and, in principle, in other forms, is suggested by writing (6.14) in the form

$$w(z) = e^{\lambda z + \log z^{\sigma} + \log\{g(z)\}}$$

$$\sim e^{\lambda z + \sigma \log z + \log\,(\alpha_0 + \alpha_1/z + \cdots)} \qquad \text{as } z \to \infty$$

$$= e^{\lambda z + \sigma \log z + \log \alpha_0 + \log\,(1 + \alpha_1/\alpha_0 z + \cdots)} \qquad \text{as } z \to \infty$$

$$= \alpha_0\, e^{\lambda z + \sigma \log z + \alpha_1/\alpha_0 z + O(z^{-2})} \qquad \text{as } z \to \infty.$$

The expansion in the exponent of the exponential, namely

$$\lambda z + \sigma \log z + \frac{\alpha_1}{\alpha_0} \frac{1}{z} + O\left(\frac{1}{z^2}\right)$$

is an asymptotic one as $z \to \infty$. A more fundamental and general approach, which includes the above one, is to look for solutions of (6.4) by considering

$$w(z) \sim e^{\phi_0(z) + \phi_1(z) + \cdots} \text{ as } z \to \infty, \quad (6.29)$$

where $\{\phi_n(z)\}$, $n = 0, 1, \ldots$, is an asymptotic sequence as $z \to \infty$. We now make the assumption that the $\phi_n(z)$ can be differentiated twice and that the resulting series $\sum_{n=0}^{\infty} \phi_n''$ and $\sum_{n=0}^{\infty} \phi_n'$ are still asymptotic. We proceed on this basis and justify the assumption *a posteriori*. This is a necessary step in this method. We now substitute (6.29) into

(6.4), using (6.13) for $f(z)$ for illustrative purposes. This gives, on cancelling the exponentials,

$$\phi_0'' + \phi_1'' + \ldots + (\phi_0' + \phi_1' + \ldots)^2 + a_0 + \frac{a_1}{z} + \ldots \sim 0, \quad (6.30)$$

which determines the ϕ_n, $n = 0, 1, \ldots$, by successively applying the asymptotic limit $z \to \infty$. The first non-zero ϕ_n is found by a simple logical process. The leading term in (6.30), which is known, comes from $f(z)$ and is here a_0, which is an $O(1)$ term. The leading term (or terms) from $(\phi_0'' + \phi_1'' + \ldots) + (\phi_0' + \phi_1' + \ldots)^2$ must cancel with this constant a_0: this determines ϕ_0. Since ϕ_0 occurs only as a derivative ϕ_0'' and $\phi_0'^2$, we must have ϕ_0 equal to some positive power of z, which implies that, asymptotically as $z \to \infty$, $\phi_0'' = o(\phi_0'^2)$,† which in turn implies that the leading term must be $\phi_0'^2$. Thus the $O(1)$ terms in (6.30) give

$$\phi_0'^2 + a_0 \sim 0, \quad (6.31)$$

the solution of which is

$$\phi_0(z) = \lambda z + O(1) \sim \lambda z, \quad \lambda = \pm i a_0^{\frac{1}{2}}. \quad (6.32)$$

Once the first term in the ϕ_n-series is found, higher order ones are easily obtained, as we see below. Using (6.32), which implies $\phi_0'' \equiv 0$, in (6.30) and equating to zero the next asymptotic terms, namely $O(z^{-1})$ here, we get

$$2\phi_0' \, \phi_1' + a_1/z \sim 0$$

which gives, using (6.32),

$$\phi_1(z) = \sigma \log z + O(1) \sim \sigma \log z, \quad \sigma = -\frac{a_1}{2\lambda}. \quad (6.33)$$

Using (6.33) and (6.32) in (6.30), we now get for the $O(z^{-2})$ terms

$$\phi_1'' + \phi_1'^2 + 2\phi_0' \, \phi_2' + a_2/z^2 \sim 0,$$

which gives

$$\phi_2(z) \sim \frac{1}{z} \frac{1}{2\lambda} \left(\frac{a_1}{2\lambda} + \frac{a_1^2}{4\lambda^2} + a_2 \right). \quad (6.34)$$

It is clear that the $O(z^{-n})$ terms will give $\phi_n(z)$, $n \geqslant 3$, and further, that $\phi_n(z)$ will be $O(z^{-n+1})$. The *a posteriori* justification for the method requires that $\sum\limits_{n=0}^{\infty} \phi_n''$ be asymptotic, which is the case, being a simple asymptotic power series as $z \to \infty$.

† It is easily shown (exercise 3) that if $f(z) = O(z^a)$ as $z \to \infty$ then $\phi_0'' = o(\phi_0'^2)$ if $a > -2$.

If we now substitute (6.32), (6.33), and (6.34) into (6.29), we finally get, on grouping all the constants of integration into one constant, chosen for convenience as $\log \alpha_0$,

$$w(z) \sim e^{\lambda z + \sigma \log z + \log \alpha_0 + \phi_2(z) + \cdots} \text{ as } z \to \infty$$

$$= \alpha_0 \, z^\sigma \, e^{\lambda z}(1 + \phi_2(z) + \ldots)$$

$$= e^{\lambda z} \, z^\sigma \left\{ \alpha_0 + \frac{1}{z} \frac{\alpha_0}{2\lambda} \left(\frac{a_1}{2\lambda} + \frac{a_1^2}{4\lambda^2} + a_2 \right) + O\left(\frac{1}{z^2} \right) \right\},$$

$$(6.35)$$

with λ, from (6.32), being the same as in (6.16) and α_0 an arbitrary constant as before. The solution (6.35) is exactly the same as (6.14) with (6.16).

As another example of this more general method, we now find the asymptotic approximations as $z \to \infty$ of what are called *parabolic cylinder functions*, which are solutions of

$$w'' - z^2 \, w = 0, \tag{6.36}$$

an equation for which the form (6.14) is not immediately suitable. (With the solutions given below, it is clear how (6.14) could be modified to cope with it.) Substituting (6.29) into (6.36), we get for (6.30) in this case

$$\phi_0'' + \phi_1'' + \ldots + \phi_0'^2 + 2\phi_0' \, \phi_1' + \phi_1'^2 + 2\phi_0' \, \phi_2' + \ldots - z^2 \sim 0. \tag{6.37}$$

Since the dominant term in $f(z)$ is here $-z^2$, we must again have ϕ_0 equal to some positive power of z, which again implies that $\phi_0'' = o(\phi_0'^2)$. Thus the dominant term is $\phi_0'^2$, which must contribute a term $O(z^2)$ to cancel the $-z^2$. We must thus try $\phi_0 = O(z^2)$ which, of course, verifies that $\phi_0'' = O(1) = o(\phi_0'^2)$. The dominant terms in (6.37), namely $O(z^2)$, give therefore

$$\phi_0'^2 - z^2 \sim 0 \Rightarrow \phi_0(z) \sim \pm z^2/2 \text{ as } z \to \infty, \tag{6.38}$$

which gives $\phi_0' \sim \pm 1$, which will be taken into account in the $O(1)$ terms. The second-order terms in (6.37), those $O(z)$ here, now give, using (6.38),

$$\phi_1'' + 2(\pm z) \, \phi_1' + \phi_1'^2 \sim 0 \Rightarrow \phi_1(z) \equiv 0 \tag{6 39}$$

as the appropriate solution. The next order terms, those $O(1)$ in (6.37) are, using (6.38) and (6.39),

$$\phi_0'' + 2\phi_0' \, \phi_2' \sim 0 \Rightarrow (\pm 1) + 2(\pm z) \, \phi_2' \sim 0$$

$$\Rightarrow \phi_2(z) \sim -\tfrac{1}{2} \log z. \tag{6.40}$$

The next two sets of terms, namely those $O(z^{-1})$ and $O(z^{-2})$, give respectively

$$
\left.
\begin{aligned}
2\phi_0' \, \phi_3' \sim 0 &\Rightarrow \phi_3(z) \equiv 0, \\[2mm]
\phi_2'' + \phi_2'^2 + 2\phi_0' \, \phi_4' \sim 0 &\Rightarrow \phi_4(z) \sim \pm\frac{3}{16}\frac{1}{z^2}.
\end{aligned}
\right\}
\tag{6.41}
$$

The asymptotic solution is now given by (6.29) with $\phi_0, \phi_1, \phi_2, \phi_3,$ $\phi_4, \ldots$ from (6.38)–(6.41). We see that the series $\sum\limits_{n=0}^{\infty} \phi_n''$ is asymptotic, and hence so is $\sum\limits_{n=0}^{\infty} \phi_n$, which is the necessary *a posteriori* justification required for the method. The two solutions of (6.36), say $w_1(z)$ and $w_2(z)$, are thus given by

$$
\left.
\begin{aligned}
w_1(z) &\sim A \ e^{z^2/2} \ z^{-\frac{1}{2}}\left\{1+\frac{3}{16}\frac{1}{z^2}+O\left(\frac{1}{z^4}\right)\right\} \\[3mm]
w_2(z) &\sim B \ e^{-z^2/2} z^{-\frac{1}{2}}\left\{1-\frac{3}{16}\frac{1}{z^2}+O\left(\frac{1}{z^4}\right)\right\}
\end{aligned}
\right\}
\text{ as } z \to \infty, \quad (6.42)
$$

where A and B are arbitrary constants. Noting the exponential form of these solutions in (6.42), it is now clear how we could obtain them by using a modified form of (6.14), namely, by looking for solutions in the form

$$
w = e^{\lambda z^2} z^{\sigma}\left(\alpha_0 + \frac{\alpha_1}{z} + \ldots\right).
$$

The dominant term in (6.30) which had to be cancelled by the dominant term in $\phi_0'' + \ldots + (\phi_0' + \ldots)^2$ was $O(1)$, while in (6.37) it was $O(z^2)$. In these two cases we showed that ϕ_0 had to be equal to z raised to some positive power, which implied that $\phi_0'' = o(\phi_0'^2)$ as $z \to \infty$, and the dominant term was thus $\phi_0'^2$. Quite generally if the dominant term in the asymptotic expansion (6.13) of $f(z)$ in (6.4) is $O(z^a)$, where a is real, we have the result (exercise 3) that $\phi_0'' = o(\phi_0'^2)$ if $a > -2$, $\phi_0'' = O(\phi_0'^2)$ if $a = -2$, the regular singularity case, and $\phi_0'^2 = o(\phi_0'')$ if $a < -2$.

Various proofs, refinements, and extensions of the first of the above methods are given in some of the books mentioned in the bibliography. The basic ideas however are essentially the same as those above. One important extension is discussed in the next section, namely that referred to as the WKB-method. The second of the methods given above, which is the more important of the two, is also extended in the following section: it is an important topic in singular perturbation theory.

Exercises

1. Find, by the first of the two main methods of this section (that is, use (6.14)), two asymptotic solutions as $z \to \infty$ of

   (i) $w'' + \dfrac{1}{z} w = 0$;

   (ii) $w'' + 2w' + \dfrac{2}{z} w = 0$;

   (iii) $w'' + \left(-\dfrac{1}{4} + \dfrac{\gamma}{z} + \dfrac{\frac{1}{4} - v^2}{z^2} \right) w = 0$, γ, v real.

2. By the second of the two methods in this section (that is, use (6.29)), find the first few terms of the asymptotic solutions as $z \to \infty$ of

   (i) $w'' - \dfrac{1}{z} w = 0$;

   (ii) $w'' + z^2 w = 0$;

   (iii) $w'' + \dfrac{1}{z+1} w = 0$;

   (iv) $w'' + (1 + z^3)w = 0$;

   (v) $w'' + z^2 w' + zw = 0$;

   (vi) $w'' - z(z+2)w = 0$.

3. If $w'' - z^a w = 0$, show that, on writing $w(z) \sim e^{\phi_0 + \phi_1 + \cdots}$, where $\{\phi_n(z)\}$, $n = 0, 1, 2, \ldots$, is an asymptotic sequence as $z \to \infty$, the differential equation becomes $\phi_0'' + \phi_1'' + \ldots + (\phi_0' + \phi_1' + \ldots)^2 - z^a \sim 0$, and that the dominant terms are $\phi_0'^2 - z^a \sim 0$ for $a > -2$, $\phi_0'' + \phi_0'^2 - z^a \sim 0$ for $a = -2$ ($z = \infty$ is a regular singularity in this case), and $\phi_0'' - z^a \sim 0$ for $a < -2$, and hence find the leading term in the asymptotic solutions in each case.

4. Find the first few terms in the asymptotic solutions as $z \to \infty$ of the following:

   (i) $w'' - \dfrac{1}{z^{\frac{1}{2}}} w = 0$;

   (ii) $w'' - \dfrac{z}{(z+1)^4} w = 0$.

6.2. Asymptotic solutions with a large or small parameter (WKB-method)

The second class of equations we consider is an important one, namely that in which the functions p and q in (6.1), and hence f in (6.4), are also dependent on a parameter λ, which may be large or small; in the latter case we conventionally denote it by ε. We first discuss the asymptotic methods in which $\lambda \to \infty$. We shall consider equations in which the independent variable is real and so we shall use x instead of z. As before, we transform (6.1) so that it is of the form (6.4), where now $f = f(x; \lambda)$ and $w = w(x; \lambda)$ are functions of x and λ. We thus consider

$$w'' + f(x; \lambda)\, w = 0, \tag{6.43}$$

where the prime denotes differentiation with respect to x. We shall find asymptotic approximations to the solutions $w(x; \lambda)$ of (6.43) as $\lambda \to \infty$ when $f(x; \lambda)$ possesses an asymptotic expansion as $\lambda \to \infty$ and, in this section, $f(x; \lambda)$ not zero in the range of x of interest. The range of x depends, as we shall see, on what questions we want answered in the given problems. Unless stated otherwise, which we do at various places below, we shall consider x to be $O(1)$.

Equation (6.43), with $f(x; \lambda) = \lambda^2 \phi_0(x)$, was first studied by Liouville[†]; and such an equation is often called, and rightly so, a *Liouville equation*. In the case when $f(x; \lambda) = \lambda^2 \phi_0(x) + \lambda\, \phi_1(x) + \phi_2(x)$, where $\phi_0(x)$, $\phi_1(x)$, and $\phi_2(x)$ are continuous, the equation was studied independently in the 1920s by Wentzel, Kramers, and Brillouin, which accounts for the WKB name for their procedure which is a special case (exercise 2) of the more comprehensive method discussed here. The method of solution they used, however, is in essence much older and was developed by Liouville, Green[‡], and Horn[§]. The description WKB-method is now retained only as a means of identification of the method.

The approach we use here is an immediate and straightforward extension of the second and more important of the two methods developed in the last section, §6.1, namely that with (6.29) as the starting point. The method here is easy to use, more fundamental,

† Liouville (1837). *J. des Math.* **2**, 22–25.

‡ Green, G. (1837). *Camb. Phil. Trans.* **6**, 457–462.

§ Horn, J. (1899). *Math. Ann.* **52**, 271–292, 340–362 and (1903). *Archiv. Math. Phys.* (3) **4**, 213–230.

and has much wider applicability than the standard WKB-procedure (see exercise 2) which is based on an extension of the first of the two methods in §6.1. The method below is also in keeping with the philosophy of singular perturbation theory of which it forms a part. It also serves to introduce the important concept of matched asymptotic expansions. In this section we also briefly introduce a few singular perturbation ideas and discuss one method for solving a class of singular perturbation problems which occur frequently in practice.

To illustrate the main method, we consider a less general $f(x; \lambda)$ than we could use, as will be clear in what follows. Specifically, let us suppose that $f(x; \lambda)$ in (6.43) has an asymptotic expansion of the form

$$
\left.
\begin{aligned}
f(x; \lambda) &\sim \lambda^2 \, \phi_0(x) + \lambda \, \phi_1(x) + \phi_2(x) + \ldots \\
&= \sum_{n=0}^{\infty} \lambda^{2-n} \, \phi_n(x)
\end{aligned}
\right\} \text{ as } \lambda \to \infty, \quad (6.44)
$$

where the $\phi_n(x)$, $n = 0, 1, \ldots$, are continuous twice-differentiable functions of x. For the moment we shall assume $\phi_0(x) \neq 0$ for any x in the range in question, and further, that the $\phi_n(x)$ are nowhere large as $O(\lambda)$: we relax the latter condition later. As we shall see, this latter condition is crucial. In §6.3 we discuss the situation where $\phi_0(x) = 0$ at some point in the range of x of interest.

In the last section we found solutions of (6.3), where f had an asymptotic expansion as $z \to \infty$, by considering w in the form (6.29), that is, an exponential with the exponent an asymptotic expansion for $z \to \infty$. It is therefore natural here to expect to find asymptotic solutions to (6.43) by looking for them as exponentials again, but where the exponent is now an asymptotic expansion as the parameter $\lambda \to \infty$, that is, of the form

$$
w(x; \lambda) \sim \exp\{g_0(\lambda)\psi_0(x) + g_1(\lambda)\psi_1(x) + \ldots\} = \exp\left\{ \sum_{n=0}^{\infty} g_n(\lambda) \, \psi_n(x) \right\}
$$

$$
\text{as } \lambda \to \infty, \quad (6.45)
$$

where $\psi_n(x)$, $n = 0, 1, \ldots$, and the sequence $\{g_n(\lambda)\}$, $n = 0, 1, \ldots$, which is an asymptotic one as $\lambda \to \infty$, have to be found. In view of the specific form, (6.44), which starts with a term $O(\lambda^2)$, we expect the $\{g_n(\lambda)\}$ here to be an asymptotic sequence in λ: this, as we show, turns out to be the case. If the first term in (6.44) is $O(\lambda)$, the sequence $\{g_n(\lambda)\}$ is an asymptotic one in half-powers of λ, as we see later. We now substitute (6.45), with the form (6.44) for f, into (6.43) just as we did in §6.1 and equate like asymptotic terms. Equation (6.43) becomes, on cancelling out the exponential in each term,

$$g_0(\lambda)\,\psi_0'' + g_1(\lambda)\,\psi_1'' + \ldots + \{g_0(\lambda)\,\psi_0' + g_1(\lambda)\,\psi_1' + \ldots\}^2 +$$
$$+ \lambda^2\,\phi_0(x) + \lambda\,\phi_1(x) + \ldots \sim 0. \tag{6.46}$$

Equating like asymptotic terms in (6.46) (here it is even simpler than in §6.1), we get

$$O(\lambda^2) \quad : \quad g_0^2(\lambda)\,\psi_0'^2(x) + \lambda^2\,\phi_0(x) \sim 0 \tag{6.47}$$

and so

$$g_0(\lambda) = \lambda, \quad \psi_0'(x) = \pm i\,\phi_0^{\frac{1}{2}}(x) \Rightarrow \psi_0(x) = \pm i \int^x \phi_0^{\frac{1}{2}}(x')\,dx', \tag{6.48}$$

where the integration constant is absorbed in the indefinite integral. Proceeding to the next term in (6.46), using (6.48), we get

$$O(\lambda) \quad : \quad g_0(\lambda)\,\psi_0''(x) + 2g_0(\lambda)\,g_1(\lambda)\,\psi_0'(x)\,\psi_1'(x) + \lambda\,\phi_1(x) \sim 0, \tag{6.49}$$

which gives

$$g_1(\lambda) = 1, \quad \psi_1'(x) = -\frac{\psi_0''(x)}{2\psi_0'(x)} - \frac{1}{2}\frac{\phi_1(x)}{\psi_0'(x)}$$

$$\Rightarrow \psi_1(x) = -\frac{1}{2}\log\psi_0'(x) - \frac{1}{2}\int^x \frac{\phi_1(x')}{\psi_0'(x')}\,dx'$$

$$= -\frac{1}{2}\log\{\pm i\,\phi_0^{\frac{1}{2}}(x)\} \mp \frac{1}{2i}\int^x \frac{\phi_1(x')}{\phi_0^{\frac{1}{2}}(x')}\,dx',$$

and so, incorporating $\log(\pm i)$ into the constant in the indefinite integral,

$$g_1(\lambda) = 1, \quad \psi_1(x) = \log\phi_0^{-\frac{1}{4}}(x) \pm \frac{i}{2}\int^x \frac{\phi_1(x')}{\phi_0^{\frac{1}{2}}(x')}\,dx'. \tag{6.50}$$

If $\phi_0(x)$ were equal to zero at some x in the range of interest, it would be a singularity of $\psi_1(x)$, in which case the method would not be valid. We thus see here the necessity for the restriction that $\phi_0(x) \neq 0$ for all x in the range in question. We come back to this point below (and in detail in §6.3) in one of the illustrative examples. The $O(1)$ terms give, from (6.46), with (6.48) and (6.50),

$$O(1) \quad : \quad g_1(\lambda)\,\psi_1''(x) + g_1^2(\lambda)\psi_1'^2(x) + 2g_0(\lambda)\,g_2(\lambda)\,\psi_0'(x)\,\psi_2'(x) \sim 0, \tag{6.51}$$

which gives $g_2(\lambda) = 1/\lambda$ and $\psi_2(x)$. With the specific form (6.44) for $f(x; \lambda)$ it is now clear, from (6.48), (6.50), and (6.51), that

$$\{g_n(\lambda)\} = \{\lambda^{1-n}\}, \quad n = 0, 1, \dots .. \tag{6.52}$$

Using the results from (6.48), (6.49), and (6.52) in (6.45), we finally get the asymptotic solutions to (6.43), with f as in (6.44), as

$$w(x; \lambda) \sim \exp\left(\pm i\lambda \int^{x} \phi_0^{\frac{1}{2}}(x')dx' + \log \phi_0^{-1/4}(x) \pm \right.$$

$$\left. \pm \frac{i}{2} \int^{x} \frac{\phi_1(x')}{\phi_0^{\frac{1}{2}}(x')} \, dx' + O\!\left(\frac{1}{\lambda}\right) \right).$$

And so the two solutions are given asymptotically by

$$w(x; \lambda) \sim \frac{1}{\phi_0^{\frac{1}{4}}(x)} \exp\left(\pm i \int^{x} \left\{ \lambda\phi_0^{\frac{1}{2}}(x') + \frac{1}{2}\frac{\phi_1(x')}{\phi_0^{\frac{1}{2}}(x')} \right\} dx' \right) \times$$

$$\times \left\{ 1 + O\!\left(\frac{1}{\lambda}\right) \right\}, \tag{6.53}$$

as long as $\phi_0(x) \neq 0$ in the x-range in question.

Liouville's equation, which frequently occurs in practical problems, is, as mentioned before, (6.43) with $f(x; \lambda) = \lambda^2\,\phi_0(x)$, namely,

$$w'' + \lambda^2\,\phi_0(x)\,w = 0. \tag{6.54}$$

The asymptotic solutions as $\lambda \to \infty$ are given immediately by (6.53), with $\phi_1(x) \equiv 0$, as

$$w(x; \lambda) \sim \frac{1}{\phi_0^{\frac{1}{4}}(x)} \exp\left\{ \pm i\lambda \int^{x} \phi_0^{\frac{1}{2}}(x')dx') + O\!\left(\frac{1}{\lambda}\right) \right\}, \tag{6.55}$$

as long as $\phi_0(x) \neq 0$.

If, in (6.44), $\phi_0(x) \equiv 0$, $\phi_1(x) \neq 0$ in the range of interest in x, the procedure (6.46) shows immediately that the first terms are $O(\lambda)$, and in place of (6.47) we have

$$O(\lambda) \quad : \quad g_0^2(\lambda)\,\psi_0'^2(x) + \lambda\,\phi_1(x) \sim 0, \tag{6.56}$$

which implies that

$$g_0(\lambda) = \lambda^{\frac{1}{2}}, \quad \psi_0(x) = \pm i \int^x \phi_1^{\frac{1}{2}}(x')\, dx'.$$

In this case in place of (6.52),

$$\{g_n(\lambda)\} = \{\lambda^{(1-n)/2}\}, \quad n = 0, 1, \ldots . \tag{6.57}$$

and the solutions are

$$w(x; \lambda) \sim \frac{1}{\phi_1^{\frac{1}{4}}(x)} \exp\left(\pm i \int^x \left\{\lambda^{\frac{1}{2}}\phi_1^{\frac{1}{2}}(x') + \frac{\phi_2(x')}{2\phi_1^{\frac{1}{2}}(x')}\right\} dx'\right)\left\{1 + O\left(\frac{1}{\lambda^{\frac{1}{2}}}\right)\right\}. \tag{6.58}$$

Note that this procedure, namely that of looking for solutions in the form (6.45), is *not* dependent on $f(x; \lambda)$ having an expansion of the specific form (6.44); all that is required is that it has some asymptotic expansion or consists of a finite number of terms or, of course, that the λ and x variations in $f(x; \lambda)$ are separable, that is, $f(x; \lambda) = F(x) G(\lambda)$. In this latter case we simply have Liouville's equation with $G(\lambda)$ playing the role of λ^2. When the series (6.44) terminates at $n = 2$, we have the case which is customarily treated by the standard WKB-procedure (exercise 2): (6.53) gives the result for this to $O(1)$.

One very important difference between the procedure discussed here and that in §6.1 is that in this section we can obtain asymptotic solutions to given boundary value problems. Among other things, we demonstrate this in some of the following illustrative examples.

As a first example, let us consider the boundary value problem given by

$$w'' + (\lambda + x) w = 0, \quad w(0; \lambda) = a, \quad w'(0; \lambda) = b,$$
$$x \geqslant 0, \quad \lambda \to \infty, \tag{6.59}$$

where a and b are given constants. (This is, in fact, an Airy equation, one form of the solutions of which are, from §3.2 and exercise 4 there, the Airy functions Ai $(-\lambda - x)$ and Bi $(-\lambda - x)$.) We wish to find the asymptotic form of the solution as $\lambda \to \infty$. The range of x is important since if x becomes large, $O(\lambda)$ say, the analysis is, as we see below, necessarily different to that when x is always $O(1)$. In the first instance let us consider (6.59) for $x = O(1)$ and so for (6.44)

$$f(x; \lambda) = \lambda + x \Rightarrow \phi_0(x) \equiv 0, \quad \phi_1(x) = 1, \quad \phi_2(x) = x,$$
$$\phi_n(x) \equiv 0, \quad n \geqslant 3. \tag{6.60}$$

Substituting (6.45) into the equation (6.59), we get, for (6.46), using (6.60),

$$g_0(\lambda)\,\psi_0'' + g_1(\lambda)\,\psi_1'' + \ldots + \{g_0(\lambda)\,\psi_0' + g_1(\lambda)\,\psi_1' + \ldots\}^2 + \lambda + x \sim 0.$$
(6.61)

The largest order terms, here $O(\lambda)$, give (compare with (6.56))

$$g_0^2(\lambda)\,\{\psi_0'(x)\}^2 + \lambda \sim 0,$$

which implies that

$$g_0(\lambda) = \lambda^{\frac{1}{2}}, \quad \psi_0(x) = \pm ix + constant.$$
(6.62)

The next terms in (6.61), namely $O(\lambda^{\frac{1}{2}})$, give

$$2g_0(\lambda)\,g_1(\lambda)\,\psi_0'\,\psi_1' \sim 0 \Rightarrow \psi_1(x) \equiv 0.$$
(6.63)

The $O(1)$ terms in (6.61) now give, using (6.63),

$$2g_0(\lambda)\,\psi_0'\,g_2(\lambda)\,\psi_2' + x \sim 0,$$
(6.64)

that is,

$$\pm 2\lambda^{\frac{1}{2}}\,i\,g_2(\lambda)\,\psi_2'(x) + x \sim 0,$$

and so

$$g_2(\lambda) = \lambda^{-\frac{1}{2}}, \quad \psi_2(x) = \pm\frac{ix^2}{4} + constant.$$
(6.65)

The $O(\lambda^{-\frac{1}{2}})$ terms are next with

$$g_2(\lambda)\,\psi_2'' + 2g_0(\lambda)\,\psi_0'\,g_3(\lambda)\,\psi_3' \sim 0$$

which is

$$\lambda^{-\frac{1}{2}}\left(\pm\frac{i}{2}\right) + 2\lambda^{\frac{1}{2}}\,(\pm i)\,g_3(\lambda)\,\psi_3' \sim 0,$$

which gives

$$g_3(\lambda) = \lambda^{-1}, \quad \psi_3(x) = -\frac{x}{4} + constant.$$
(6.66)

Higher order terms are obtained from (6.61) by proceeding in the same way, for example,

$$g_4(\lambda) = \lambda^{-\frac{3}{2}}, \quad \psi_4(x) = \pm\frac{ix^3}{24}.$$

It is easily checked that here, with $g_1(\lambda)$ undetermined but not required,

$$g_0(\lambda) = \lambda^{\frac{1}{2}}, \quad g_2(\lambda) = \lambda^{-\frac{1}{2}}, \quad g_n(\lambda) = \lambda^{(1-n)/2}, \quad n \geqslant 2.$$

The two asymptotic solutions, w_1 and w_2 say, are now given by (6.45), using (6.62), (6.63), (6.65), and (6.66), as

$$\left. \begin{aligned} w_1(x; \lambda) &\sim \exp\left(x\left\{i\lambda^{\frac{1}{2}}+\frac{ix}{4\lambda^{\frac{1}{2}}}-\frac{1}{4\lambda}+O\left(\frac{ix^2}{\lambda^{\frac{3}{2}}}\right)\right\}\right) \\[2mm] w_2(x; \lambda) &\sim \exp\left(-x\left\{i\lambda^{\frac{1}{2}}+\frac{ix}{4\lambda^{\frac{1}{2}}}+\frac{1}{4\lambda}+O\left(\frac{ix^2}{\lambda^{\frac{3}{2}}}\right)\right\}\right) \end{aligned} \right\} x = O(1), \lambda \to \infty.$$

(6.67)

The solution to the specific boundary-value problem (6.59) is now given by

$$w(x; \lambda) = Aw_1(x; \lambda)+Bw_2(x; \lambda),$$ (6.68)

where A and B, using (6.59), must satisfy

$$A+B = a, \quad \left(i\lambda^{\frac{1}{2}}-\frac{1}{4\lambda}+\ldots\right)A-\left(i\lambda^{\frac{1}{2}}+\frac{1}{4\lambda}+\ldots\right)B = b.$$

Solving these for A and B and substituting back into (6.68), we get, as the asymptotic solution to the boundary-value problem (6.59),

$$\begin{aligned} w(x; \lambda) \sim e^{-(x/4\lambda)}&\left\{a \cos x\left(\lambda^{\frac{1}{2}}+\frac{x}{4\lambda^{\frac{1}{2}}}+O\left(\frac{x^2}{\lambda^{\frac{3}{2}}}\right)\right)+\right. \\[2mm] &+\lambda^{-\frac{1}{2}}\left(b+\frac{a}{4\lambda}\right)\sin x\left(\lambda^{\frac{1}{2}}+\frac{x}{4\lambda^{\frac{1}{2}}}+O\left(\frac{x^2}{\lambda^{\frac{3}{2}}}\right)\right)+ \\[2mm] &\left.+\ldots\right\}, \quad x = O(1), \quad \lambda \to \infty, \end{aligned}$$

which, on expanding $e^{-(x/4\lambda)}$, gives, since with $x = O(1)$ $x/\lambda \ll 1$,

$$\begin{aligned} w(x; \lambda) \sim a&\left(1-\frac{x}{4\lambda}\right)\cos x\left(\lambda^{\frac{1}{2}}+\frac{x}{4\lambda^{\frac{1}{2}}}+\ldots\right)+ \\[2mm] &+\lambda^{-\frac{1}{2}}\left(b+\frac{a}{4\lambda}\right)\sin x\left(\lambda^{\frac{1}{2}}+\frac{x}{4\lambda^{\frac{1}{2}}}+\ldots\right)+ \\[2mm] &+O\left(\frac{1}{\lambda^{\frac{3}{2}}}\right), \quad x = O(1), \quad \lambda \to \infty. \end{aligned}$$

(6.69)

Suppose now we would like to have the asymptotic solution to (6.59) for a range including x large, say $O(\lambda)$. The solutions w_1 and w_2 in (6.67) and w in (6.69) clearly do not hold when $x = O(\lambda)$ since the exponents are no longer asymptotic series as $\lambda \to \infty$ in such a case. Of course, what we would really like to find is a solution which is a *uniformly valid asymptotic solution* to the problem which holds for all $x \geqslant 0$ of interest. Generally this is the aim of all asymptotic methods. Singular perturbation theory, among other things, is directed towards providing such solutions by what is called a *matched asymptotic procedure*. This is essentially a process whereby asymptotic solutions, valid for different ranges of the parameter or variable in question, are found in such a way that one form of the solution at one limit of its validity joins or matches onto another form of the solution at one limit of its validity. We touch on this matching procedure below and in §6.2. The solution to (6.59) in terms of Airy functions is a uniformly valid solution for all values of λ and x.

Let us now consider the problem (6.59) for the range of x from $x = 0$ to $x = O(\lambda)$. We reduce the problem to one in which the independent variable is $O(1)$ by introducing

$$y = \frac{x}{\lambda}, \quad W(y; \lambda) = w(x; \lambda), \quad w' = \frac{dw}{dx} = \frac{1}{\lambda}\frac{dW}{dy} = \frac{1}{\lambda}W',$$

in which case (6.59) becomes

$$W'' + \lambda^3(1+y)\,W = 0, \quad W(0; \lambda) = a, \quad W'(0; \lambda) = \lambda b.$$
$$(6.70)$$

The asymptotic solutions of this equation are now obtained in the usual way by writing W in the form (6.45) and evaluating the sequence $\{g_n(\lambda)\}$, $n = 0, 1, 2, \ldots$, and, in this case, $\psi_n(y)$, $n = 0, 1, \ldots$. Equation (6.70) is a Liouville equation (6.54) with λ^3, y, W, and $1+y$ in place of λ^2, x, w, and ϕ_0 respectively, and the asymptotic solutions, W_1 and W_2 say, are given by (6.55) with the appropriate substitutions as

$$\frac{W_1(y; \lambda)}{W_2(y; \lambda)} \sim \frac{1}{(1+y)^{\frac{1}{4}}} \exp\left(\pm i\lambda^{\frac{3}{2}}\int^y (1+y')^{\frac{1}{2}}dy' + O\left(\frac{1}{\lambda^{\frac{3}{2}}}\right)\right) \quad (6.71)$$

which in terms of $x(= \lambda y)$ and $w(= W)$ are

$$\frac{w_1(x; \lambda)}{w_2(x; \lambda)} \sim \frac{1}{\left(1+\frac{x}{\lambda}\right)^{\frac{1}{4}}} \exp\left(\pm\frac{2i}{3}\lambda^{\frac{3}{2}}\left(1+\frac{x}{\lambda}\right)^{\frac{3}{2}} + O\left(\frac{1}{\lambda^{\frac{3}{2}}}\right)\right)$$

$$0 \leqslant x \leqslant O(\lambda), \quad \lambda \to \infty. \quad (6.72)$$

Using (6.71) in (6.70), or alternatively (6.72), with (6.68), in (6.59), after a little algebra we get, by a similar process of evaluating constants A and B in (6.68), which resulted in (6.69), the solution to the boundary value problem, as

$$w(x; \lambda) \sim \frac{1}{\left(1+\frac{x}{\lambda}\right)^{\frac{1}{4}}} \left(a \cos\left[\frac{2}{3} \lambda^{\frac{3}{2}} \left\{\left(1+\frac{x}{\lambda}\right)^{\frac{3}{2}} - 1\right\} + O\left(\frac{1}{\lambda^{\frac{3}{2}}}\right)\right] + \right.$$

$$+ \frac{1}{\lambda^{\frac{1}{2}}}\left(b + \frac{a}{4\lambda}\right) \sin\left[\frac{2}{3} \lambda^{\frac{3}{2}} \left\{\left(1+\frac{x}{\lambda}\right)^{\frac{3}{2}} - 1\right\} + O\left(\frac{1}{\lambda^{\frac{3}{2}}}\right)\right] +$$

$$\left. + O\left(\frac{1}{\lambda^{\frac{3}{2}}}\right)\right), \quad x \leqq O(\lambda), \quad \lambda \to \infty. \tag{6.73}$$

The solution (6.73) holds for *all* x up to $O(\lambda)$. To see that (6.73) includes the $x = O(1)$ form of the solution (6.69), we simply consider (6.73) for $x = O(1)$, in which case, $x/\lambda \ll 1$ and

$$\left. \begin{array}{l} \dfrac{2}{3} \lambda^{\frac{3}{2}} \left\{\left(1+\dfrac{x}{\lambda}\right)^{\frac{3}{2}} - 1\right\} = x\left(\lambda^{\frac{1}{2}} + \dfrac{x}{4\lambda^{\frac{1}{2}}} + O\left(\dfrac{x^2}{\lambda^{\frac{3}{2}}}\right)\right), \\[3mm] \dfrac{1}{\left(1+\dfrac{x}{\lambda}\right)^{\frac{1}{4}}} = 1 - \dfrac{x}{4\lambda} + O\left(\dfrac{x^2}{\lambda^2}\right), \end{array} \right\}$$

use of which in (6.73) gives exactly (6.69). Thus (6.73) gives the *uniformly valid asymptotic solution* to (6.59) for all $x \geqslant 0$ up to $x = O(\lambda)$. Note from (6.73) that the solution is oscillatory in character, as would be expected, of course, from the form of (6.59), since $\lambda + x > 0$.

A final point about this problem (6.59) is that for any λ, no matter how large, as long as it is not infinite, the asymptotic solution as $x \to \infty$ must be obtained by the methods of the §6.1, in which case we cannot solve the specific boundary value problem. If we consider the class of equations in §6.1, we can use the above procedure to obtain their asymptotic solutions as $z \to \infty$ if we introduce artificially a large parameter by writing $z = \lambda Z$ and then consider $Z = O(1)$ and $\lambda \to \infty$. We still cannot solve a specific boundary value problem in this case, of course.

Suppose now we consider a problem which is related to that in (6.59) but which involves a small parameter ε which tends to zero, namely

$$w'' + (1 + \varepsilon t) w = 0, \quad w(0; \varepsilon) = c, \quad w'(0; \varepsilon) = d, \qquad (6.74)\dagger$$

where here the prime denotes differentiation with respect to the independent variable t (usually time in such equations). The problem here is to find the asymptotic solution as $\varepsilon \to 0$. Let us suppose also that we require the uniformly valid solution for all t up to say, $t = O(1/\varepsilon) \gg 1$. Because $\varepsilon \ll 1$ we might first think of trying a simple regular perturbation scheme by writing $w(t; \varepsilon)$ as a convergent series in ε, namely

$$w(t; \varepsilon) = F_0(t) + \varepsilon F_1(t) + \ldots \qquad (6.75)$$

Substituting this form into (6.74) and equating powers of ε, we obtain the $F_0(t)$, $F_1(t)$, ... successively. We shall do this to show how such a scheme does not give a uniformly valid solution for all t. Carrying out this procedure gives

$$(F_0'' + \varepsilon F_1'' + \ldots) + (1 + \varepsilon t) (F_0 + \varepsilon F_1 + \ldots) = 0.$$

On collecting like terms and successively applying the limit $\varepsilon \to 0$,

$$F_0'' + F_0 = 0, \quad F_1'' + F_1 = -tF_0, \quad F_2'' + F_2 = -tF_1, \ldots \qquad (6.76)$$

The solution $F_0(t)$ satisfying the initial conditions (6.74) is

$$F_0(t) = c \cos t + d \sin t. \qquad (6.77)$$

The problem for $F_1(t)$ is thus, from (6.76), with (6.77) satisfying the initial conditions in (6.74),

$$F_1'' + F_1 = -t(c \cos t + d \sin t), \quad F_1(0) = 0 = F_1'(0),$$

the solution of which is

$$F_1(t) = \frac{c}{4} \sin t - \frac{t}{4}\{c(\cos t + t \sin t) + d(\sin t - t \cos t)\}. \qquad (6.78)$$

The solution to $O(\varepsilon)$ is therefore

$$w(t; \varepsilon) = (c \cos t + d \sin t) + \frac{\varepsilon}{4}\{c \sin t - ct(\cos t + t \sin t) -$$

$$- dt(\sin t - t \cos t)\} + O(\varepsilon^2). \qquad (6.79)$$

† Physically this represents an oscillation with a slowly varying frequency (see also (6.93) below et seq.).

The series solution (6.75) obtained in this way is a valid asymptotic one for $\varepsilon \to 0$ *only* as long as $t = O(1)$. If t becomes large then εt and εt^2 in (6.79) are no longer $O(\varepsilon)$ and so $\varepsilon F_1(t)$ is no longer $O(\varepsilon)$, the premise on which solutions of the form (6.75), namely a convergent asymptotic series, is based. These terms, $t \cos t$, $t^2 \sin t$, and so on in (6.79) are called *secular terms* and it is their presence which invalidates the simple procedure (6.75) for $t = O(1/\varepsilon)$: the series is no longer asymptotic. Such problems as (6.74) hitherto have been most frequently solved by what is called a multi-variable or two-time expansion procedure. This specific equation (6.74), in fact, belongs to a class treated in Cole's (1968) book by a careful and necessarily non-standard application of such a two-variable expansion procedure. By the method we have developed in this section, the uniformly asymptotic solution can be trivially found by a straightforward use†
of the solution form (6.45) on rewriting (6.74) appropriately in terms of $\lambda = 1/\varepsilon$, keeping in mind that we require the solution valid for $t = O(1/\varepsilon) = O(\lambda)$. This we now do.

So as to include times $t = O(1/\varepsilon)$, we write (compare with the above in regard to (6.70))

$$\Lambda = \frac{1}{\varepsilon}, \quad Y = \varepsilon t = \frac{t}{\Lambda}, \quad W(Y; \Lambda) = w(t; \varepsilon), \qquad (6.80)$$

and (6.74) becomes

$$W'' + \Lambda^2(1+Y)W = 0, \quad W(0; \Lambda) = c, \quad W'(0; \Lambda) = \Lambda d, \tag{6.81}$$

where prime denotes differentiation with respect to Y. This is exactly the same as problem (6.70) if we associate respectively λ, y, a, and b there with $\Lambda^{\frac{2}{3}}$, Y, c and $\Lambda^{\frac{1}{3}}d$ in (6.81): these imply that

$$\lambda^{\frac{3}{2}} = \Lambda = \frac{1}{\varepsilon}, \quad \frac{x}{\lambda} = \varepsilon t, \quad \frac{b}{\lambda^{\frac{3}{2}}} + \frac{a}{4\lambda^{\frac{3}{2}}} = \frac{\Lambda^{\frac{1}{3}}d}{\Lambda^{\frac{3}{2}}} + \frac{c}{4\Lambda} = d + \frac{\varepsilon c}{4}.$$

The uniformly valid asymptotic solution is given immediately by (6.73), on writing there

$$\frac{x}{\lambda} = \varepsilon t, \quad a = c, \quad \lambda^{\frac{3}{2}} = \frac{1}{\varepsilon}, \quad \frac{1}{\lambda^{\frac{3}{2}}}\left(b + \frac{a}{4\lambda}\right) = d + \frac{\varepsilon c}{4},$$

$$w(x; \lambda) = w(t; \varepsilon),$$

† This approach was suggested to the author by Dr. J. R. Ockendon.

as

$$w(t; \varepsilon) \sim \frac{1}{(1+\varepsilon t)^{\frac{1}{4}}} \left[c \cos \frac{2}{3\varepsilon} \{(1+\varepsilon t)^{\frac{3}{2}} - 1\} + \right.$$
$$\left. + \left(d + \frac{\varepsilon c}{4} \right) \sin \frac{2}{3\varepsilon} \{(1+\varepsilon t)^{\frac{3}{2}} - 1\} \right]$$

$$\text{as } \varepsilon \to 0,\ 0 \leq t \leq O(1/\varepsilon). \qquad (6.82)\dagger$$

We have dealt at length with these examples, which incidentally are all forms of Airy's equation, since they bring out various important points of the general method and introduce the concept of uniformly valid solutions. The close connection between this method and singular perturbation theory has been demonstrated. Below we discuss its direct application in this field. As a method for treating *linear* and *weakly nonlinear* problems, normally considered by a two-variable procedure, the above procedure is considerably easier to use and also appears to cover a wider class of problems.

The most frequently occurring singular perturbation problems are those in which a small parameter multiplies the highest derivative in a differential equation: (6.81) with $\Lambda = 1/\varepsilon$ is an example of such an equation when written as $\varepsilon^2 W'' + (1+Y)W = 0$. We thus have a procedure, which is a straightforward application of the above, for finding asymptotic solutions, as $\varepsilon \to 0$, to a wide class of singular perturbation problems, namely those, for example (6.1), which are reducible to the class of equations

$$\varepsilon w'' + f(x; \varepsilon)w = 0, \qquad (6.83)$$

in which $f(x; \varepsilon)$ has an asymptotic expansion as $\varepsilon \to 0$. We also require $f(x; \varepsilon) \neq 0$ for x in the range in question. On writing $\varepsilon = 1/\lambda$, (6.83) becomes $w'' + \lambda f(x; 1/\lambda)w = 0$, which is a member of the class (6.43) and which is solved asymptotically by looking for solutions of the form (6.45). A method for solving (6.83) when

$$\left. \begin{array}{l} f(x; \varepsilon) \sim h_0(\varepsilon)\ \phi_0(x) + h_1(\varepsilon)\ \phi_1(x) + \ldots, \\[2mm] h_{n+1}(\varepsilon) = o(h_n(\varepsilon)) \text{ as } \varepsilon \to 0, \end{array} \right\} \qquad (6.84)$$

where $\phi_n(x)$, $n = 0, 1, 2, \ldots$, are twice-differentiable $O(1)$ functions of x and $\{h_n(\varepsilon)\}$ is an asymptotic sequence as $\varepsilon \to 0$, is to write

$$w(x; \varepsilon) \sim \exp (g_0(\varepsilon)\ \psi_0(x) + g_1(\varepsilon)\ \psi_1(x) + \ldots) \text{ as } \varepsilon \to 0 \quad (6.85)$$

---

† Note the effect on the oscillations due to the slowly varying frequency in (6.74) (see the footnote on page 120).

where $\{g_n(\varepsilon)\}$, an asymptotic sequence as $\varepsilon \to 0$, and $\psi_n(x)$, $n = 0, 1, 2, \ldots$, have to be determined. Even though the method is in essence a reformulation of the above $\lambda \to \infty$ procedure, we carry out the analysis here because of its considerable importance in its own right.

By way of illustration, let us find the asymptotic solutions as $\varepsilon \to 0$ of

$$
\left.
\begin{aligned}
& \varepsilon\, w'' + f(x; \varepsilon)\, w = 0, \\
& f(x; \varepsilon) \sim \phi_0(x) + \varepsilon\, \phi_1(x) + \varepsilon^2\, \phi_2(x) + \ldots,
\end{aligned}
\right\} \text{as } \varepsilon \to 0. \quad (6.86)
$$

Substituting (6.85) into (6.86), we get

$$
\varepsilon(g_0\, \psi_0'' + g_1\, \psi_1'' + \ldots) + \varepsilon(g_0\, \psi_0' + g_1\, \psi_1' + \ldots)^2 +
$$
$$
+ (\phi_0 + \varepsilon\, \phi_1 + \ldots) \sim 0. \quad (6.87)
$$

We now proceed in a comparable way to that used on (6.46), except that here we find g_n, ψ_n, $n = 0, 1, 2, \ldots$, by successively applying the limit $\varepsilon \to 0$. From (6.87) the $O(1)$ terms give

$$
O(1): \quad \varepsilon g_0^2\, \psi_0'^2 + \phi_0(x) \sim 0,
$$

which gives

$$
g_0(\varepsilon) = \frac{1}{\varepsilon^{\frac{1}{2}}}, \quad \psi_0(x) = \pm i \int^x \phi_0^{\frac{1}{2}}(x')\, \mathrm{d}x', \quad (6.88)
$$

which, on using in (6.87), gives the next terms as

$$
O(\varepsilon^{\frac{1}{2}}): \quad \varepsilon^{\frac{1}{2}}\, \psi_0'' + 2\varepsilon^{\frac{1}{2}}\, g_1(\varepsilon)\, \psi_0'\, \psi_1' \sim 0,
$$

and so

$$
g_1(\varepsilon) = 1, \quad \psi_1(x) = -\tfrac{1}{4} \log \phi_0(x). \quad (6.89)
$$

With (6.88) and (6.89) we have

$$
O(\varepsilon): \quad \varepsilon\, \psi_1'' + \varepsilon\, \psi_1'^2 + 2\varepsilon^{\frac{3}{2}}\, g_2(\varepsilon)\, \psi_0'\, \psi_2' + \varepsilon\, \phi_1(x) \sim 0,
$$
$$
(6.90)
$$

which gives $g_2(\varepsilon) = \varepsilon^{\frac{1}{2}}$ and $\psi_2(x)$ on integration. Higher-order terms are obtained in a similar way. The asymptotic solutions to (6.86) are thus, from (6.85), with (6.88)–(6.90),

$$
w(x; \varepsilon) \sim \exp\left(\pm \frac{i}{\varepsilon^{\frac{1}{2}}} \int^x \phi_0^{\frac{1}{2}}(x')\mathrm{d}x' - \tfrac{1}{4} \log \phi_0(x) + 0(\varepsilon^{\frac{1}{2}}) \right)
$$

$$
\sim \frac{1}{\phi_0^{\frac{1}{4}}(x)} \exp\left(\pm \frac{i}{\varepsilon^{\frac{1}{2}}} \int^x \phi_0^{\frac{1}{2}}(x')\mathrm{d}x' \right) \text{as } \varepsilon \to 0. \quad (6.91)
$$

An alternative form of the asymptotic solution of (6.86) is from (6.91),

$$w(x; \varepsilon) \sim \frac{1}{\phi_0^{\frac{1}{4}}(x)} \left\{ A \cos \left(\frac{1}{\varepsilon^{\frac{1}{2}}} \int^x \phi_0^{\frac{1}{2}}(x')\, dx' + O(\varepsilon^{\frac{1}{2}}) \right) + \right.$$

$$\left. + B \sin \left(\frac{1}{\varepsilon^{\frac{1}{2}}} \int^x \phi_0^{\frac{1}{2}}(x')\, dx' + O(\varepsilon^{\frac{1}{2}}) \right) \right\} \text{ as } \varepsilon \to 0, \qquad (6.92)$$

where A and B are arbitrary constants to be determined by the boundary conditions of the given problem. The procedure is clearly analogous to that above in which the parameter $\lambda \to \infty$.

The experience with the above example (6.74), for which we found a uniformly valid asymptotic solution, and the general procedure for handling equations like (6.86) suggests a general method for obtaining uniformly valid asymptotic solutions for equations, for example, like (see also (6.74) above)

$$\frac{d^2 w}{dt^2} + f(\varepsilon t; \varepsilon) w = 0, \quad \varepsilon \to 0. \qquad (6.93)$$

Practically, equation (6.93) represents oscillations with frequencies which vary slowly with time t as reflected by its appearance in εt in $f(\varepsilon t; \varepsilon)$. When $\varepsilon \equiv 0$, (6.93) has the harmonic solutions $\genfrac{}{}{0pt}{}{\sin}{\cos} \omega_0 t$, where $\omega_0^2 = f(0; 0)$. When ε is small but not zero, $f(\varepsilon t; \varepsilon)$ is a slowly varying function of t and clearly makes the frequency a function of time. The procedure for finding uniformly asymptotic solutions for $t = O(1/\varepsilon)$ is thus to introduce

$$x = \varepsilon t, \quad W(x; \varepsilon) = w(t; \varepsilon), \qquad (6.94)$$

in which case (6.93) becomes

$$\varepsilon^2 W'' + f(x; \varepsilon) W = 0, \qquad (6.95)$$

where the primes denote partial differentiation with respect to x. But (6.95) is similar to (6.86) if $f(x; \varepsilon)$ here possesses an asymptotic expansion as $\varepsilon \to 0$: it is almost always the case that it does. Suppose, for illustration only, that

$$f(x; \varepsilon) \sim \Phi_0(x) + \varepsilon \Phi_1(x) + \ldots \text{ as } \varepsilon \to 0, \qquad (6.96)$$

where $\Phi_n(x)$, $n = 0, 1, 2, \ldots$, are twice differentiable and $\Phi_0(x) \neq 0$, then the asymptotic solutions of (6.95) with (6.96) are obtained by looking for solutions in the form (6.85) in the now usual way. We immediately obtain the dominant terms (to first order they are given by (6.91) or (6.92) with ε^2 for ε) or one form of them, as

$$W(x; \varepsilon) \sim \frac{1}{\Phi_0^{\frac{1}{2}}(x)} \left\{ A \cos\left(\frac{1}{\varepsilon} \int\limits^{x} \Phi_0^{\frac{1}{2}}(x')\, dx'\right) + \right.$$

$$\left. + B \sin\left(\frac{1}{\varepsilon} \int\limits^{x} \Phi_0^{\frac{1}{2}}(x')\, dx'\right)\right\}, \tag{6.97}$$

where A and B are arbitrary constants. The solutions to (6.93) are given by (6.97) on writing $x = \varepsilon t$ and w for W. The solution to a given boundary-value problem can be found by evaluating A and B. For example, suppose for (6.93) we require,

$$w(0; \varepsilon) = a, \quad w'(0; \varepsilon) = b, \tag{6.98}$$

we get, using (6.98) on (6.97) with (6.94), (see also exercise 4),

$$w(t; \varepsilon) \sim \left\{\frac{\Phi_0(0)}{\Phi_0(\varepsilon t)}\right\}^{\frac{1}{2}} \left\{ a \cos\left(\frac{1}{\varepsilon} \int\limits_0^{\varepsilon t} \Phi_0^{\frac{1}{2}}(x')\, dx'\right) + \right.$$

$$\left. + \frac{b}{\Phi_0^{\frac{1}{2}}(0)} \sin\left(\frac{1}{\varepsilon} \int\limits_0^{\varepsilon t} \Phi_0^{\frac{1}{2}}(x')\, dx'\right)\right\} \text{ as } \varepsilon \to 0. \tag{6.99}$$

At this stage we should reiterate the point that, in using the above procedures, we require $f \neq 0$ in (6.43) and (6.83) for x in the range of interest.

The above asymptotic methods, based on seeking solutions in the form of exponentials with exponents consisting of asymptotic series, are clearly applicable to an incredibly large class of linear differential equations. We chose second-order equations only for illustration and simplicity.

As an illustration of a practical application of the technique described above, consider small-amplitude high-frequency wave motion in a non-uniform medium, the effect of which is to vary the wave speed with distance. The one-dimensional wave equation for the disturbance $Z(x, t)$ in such a situation is, in dimensionless form,

$$Z_{tt} = c^2(x)\, Z_{xx}, \tag{6.100}$$

where subscripts denote partial differentiation and $c(x) > 0$ is the wave speed, assumed given, which varies with x. Wave-like solutions of (6.100) are found by looking for $Z(x, t)$ in the form

$$Z(x, t) = \text{Rl}\left\{w(x; \omega)\, e^{i\omega t}\right\}, \tag{6.101}$$

where ω is the frequency of the oscillation. Substitution of (6.101) into (6.100) gives the equation for $w(x; \omega)$ as

$$w'' + \frac{\omega^2}{c^2(x)} w = 0, \tag{6.102}$$

which belongs to the class of equations (6.43), with (6.44) if we consider high-frequency oscillations, that is, ω large: it is in fact a Liouville equation. Comparing $\omega^2/c^2(x)$ with $f(x; \lambda)$ in (6.44), we have

$$\lambda = \omega, \quad \phi_0(x) = \frac{1}{c^2(x)}, \quad \phi_n(x) \equiv 0, \quad n \geqslant 1,$$

and the asymptotic solutions are then immediately given by (6.55) as

$$w(x; \omega) \sim c^{\frac{1}{2}}(x) \exp\left(\pm i\omega \int^x \frac{dx'}{c(x')}\right) \text{ as } \omega \to \infty. \tag{6.103}$$

Asymptotic solutions of (6.100) therefore, for high-frequency oscillations, are given by (6.101), with (6.103), as

$$Z(x, t) \sim \text{Rl}\left\{c^{\frac{1}{2}}(x) \exp\left(i\omega(t \pm \int^x \frac{dx'}{c(x')})\right)\right\} \text{ as } \omega \to \infty, \tag{6.104}$$

from which we see that $|Z(x, t)| \propto c^{\frac{1}{2}}(x)$, that is, the amplitude is proportional to the square root of the local wave speed.

As an example of the use of the above ideas when there are several independent variables, consider again small-amplitude high-frequency oscillations in a two-dimensional uniform medium. The governing equation for the disturbance $Z(x, y, t)$ is here

$$\Delta Z = Z_{xx} + Z_{yy} = c^2 Z_{tt}, \tag{6.105}$$

where c, the wave speed, is taken to be constant. High-frequency wave solutions are again found on setting (compare with (6.101))

$$Z(x, y, t) = \text{Rl}\{e^{i\omega t} w(x, y; \omega)\}, \tag{6.106}$$

where ω is the frequency which is again taken to be large. Substituting (6.106) into (6.105) gives

$$w_{xx} + w_{yy} + \lambda^2 w = 0, \quad \lambda = \omega c \gg 1. \tag{6.107}$$

We now look for asymptotic solutions of (6.107) in the form

$$w(x, y; \lambda) \sim \exp[g_0(\lambda)\psi_0(x,y) + g_1(\lambda)\psi_1(x,y) + \ldots] \text{ as } \lambda \to \infty,$$

$$\tag{6.108}$$

where $\{g_n(\lambda)\}$, $n = 0, 1, 2, \ldots$, is an asymptotic sequence as $\lambda \to \infty$ and it and the $\psi_n(x, y)$, $n = 0, 1, 2, \ldots$, have to be found. Substituting (6.108) into (6.107) gives

$$g_0 \, \Delta\psi_0 + g_1 \, \Delta\psi_1 + \ldots +$$

$$+ \{(g_0 \, \psi_{0_x} + g_0 \, \psi_{0_y}) + (g_1 \, \psi_{1_x} + g_1 \, \psi_{1_y}) + \ldots\}^2 + \lambda^2 \sim 0.$$

The first-order terms, $O(\lambda^2)$ here, give

$$g_0^2(\lambda) \, (\psi_{0_x}^2 + \psi_{0_y}^2) + \lambda^2 \sim 0$$

and so

$$g_0(\lambda) = \lambda, \quad \psi_{0_x}^2 + \psi_{0_y}^2 + 1 = 0. \tag{6.109}$$

This partial differential equation, which is the equation of geometrical optics, can be solved by characteristics: see, for example, Garabedian (1964)†. The dominant term in the asymptotic solution of (6.105) is thus, from (6.106), (6.108), and (6.109) with $\lambda = \omega c$,

$$Z(x, y, t; \omega) \sim \mathrm{Rl} \, \{\exp[i\omega t + \omega c\psi_0(x, y) + o(\omega)]\} \text{ as } \omega \to \infty,$$

$$\tag{6.110}$$

where $\psi_0(x, y)$ is the solution of (6.109).

As another practical illustration, also associated with wave phenomena, we consider *Schrödinger's equation* from quantum mechanics which, in its one-dimensional form convenient for discussion here, that is in a form like (6.43), is

$$w'' + \lambda^2 \{E(\lambda) - V(x)\} w = 0, \tag{6.111}$$

where $E(\lambda)$, an energy level, is a function of λ, and $V(x)$, a potential, is a function of x: both $E(\lambda)$ and $V(x)$ are given. With $\lambda \gg 1$ we have what is known as the *short wave-length approximation*. We wish to find the asymptotic form of the solutions $w(x; \lambda)$ of (6.111) as $\lambda \to \infty$.

To start with let us take the simple case in which $E(\lambda) = E_0$, a constant. On comparison of (6.111) with (6.43) and (6.44), we have

$$\phi_0(x) = E_0 - V(x), \quad \phi_n(x) \equiv 0, \quad n \geqslant 1, \tag{6.112}$$

and so (6.111) is again a Liouville equation. If, in the first instance, we suppose $E_0 > V(x)$ for all x of interest, $\phi_0(x)$ is *positive* and the

† Garabedian, P. R. (1964). *Partial differential equations*. Wiley, New York.

asymptotic solutions of (6.111) with (6.112) are given immediately by (6.55) as

$$w(x; \lambda) \sim \{E_0 - V(x)\}^{-\frac{1}{4}} \exp(\pm i\lambda \int^x \{E_0 - V(x')\}^{\frac{1}{2}} dx') \text{ as } \lambda \to \infty.$$

(6.113)

In this case the solution is oscillatory in character.

If we now suppose $E_0 < V(x)$, for all x of interest, $\phi_0(x)$ in (6.112) is *negative* and the asymptotic solutions from (6.55) are now

$$w(x; \lambda) \sim \{V(x) - E_0\}^{-\frac{1}{4}} \exp(\pm \lambda \int^x \{V(x') - E_0\}^{\frac{1}{2}} dx') \text{ as } \lambda \to \infty,$$

(6.114)

which is exponential rather than oscillatory in character.

This interesting difference in the asymptotic character of the solutions, namely (6.113) being oscillatory and (6.114) being exponential, is related, of course, to the sign of $\phi_0(x)$ of (6.112). For $\phi_0(x) > 0$ we expect some sort of oscillatory solution for (6.111), whereas for $\phi_0(x) < 0$ we expect a behaviour exhibiting some sort of exponential growth or decay. Clearly if there is an $x = x_c$, say, in the range of x where $V(x_c) = E_0$ and $V'(x_c) \neq 0$, $\phi_0(x)$ changes sign in the range of interest and has a zero at $x = x_c$. On one side of $x = x_c$ the solution will be like one of (6.113) or (6.114), while on the other side it will be the other of (6.113) or (6.114), with *neither* valid in the vicinity of $x = x_c$ which would be a singularity of the solutions since $\phi_0(x_c) = 0$ there. Such a point is called a *transition point*. The word transition is used because the transition from one type of solution to the other must take place in the vicinity of such a point; that is, one where $\phi_0(x) = 0$. Generally, transition points of (6.43) are where $f(x; \lambda) = 0$. Transition points are discussed below in §6.3.

Suppose now we consider $E(\lambda)$ to vary with λ and write, as an example, $E(\lambda) = E_0/\lambda$, where E_0 is a constant. Then (6.111) becomes

$$w'' - \{\lambda^2 V(x) - \lambda E_0\}w = 0,$$ (6.115)

which on comparison with (6.43) and (6.44) has

$$\phi_0(x) = -V(x), \quad \phi_1(x) = E_0, \quad \phi_n(x) \equiv 0, \quad n \geqslant 2.$$

(6.116)

In this case (6.53) is again immediately applicable with $\phi_0^{\frac{1}{2}} = \pm i V^{\frac{1}{2}}(x)$

and so,

$$w(x; \lambda) \sim \frac{1}{V^{\frac{1}{4}}(x)} \exp(\pm \int^x \{\lambda V^{\frac{1}{2}}(x') - \tfrac{1}{2}E_0 V^{-\frac{1}{2}}(x')\}dx')$$

$$\text{as } \lambda \to \infty, \quad (6.117)$$

which, although still exponential, is quite different again from the solutions (6.113) and (6.114). The solution (6.117) cannot be obtained from (6.114) on letting E_0 there simply tend to zero like $O(\lambda^{-1})$ as $\lambda \to \infty$ since in obtaining it we considered E_0 to be constant with respect to λ. When it varies like $O(1/\lambda)$ the asymptotic analysis has to be done to a higher order, as was done to obtain (6.53) and hence (6.117). The same observation applies to any λ-variation of $E(\lambda)$, of course. Each of these solutions is in fact one part of a uniformly valid asymptotic solution $w(x; \lambda)$ for $E(\lambda)$ varying from $O(1/\lambda)$ to $O(1)$. The joining together of these solutions for different $E(\lambda)$ is a problem in matched asymptotic expansions, a subject we briefly touched on above.

Exercises

1. From first principles, that is start with (6.45), find the first few terms in the asymptotic expansion as $\lambda \to \infty$ of the solutions of the following, giving in each case the sequence $\{g_n(\lambda)\}$:

   (i) $w'' + \lambda^2 xw = 0$;

   (ii) $w'' + (\lambda^2 x^2 + \lambda x + 1)w = 0$;

   (iii) $w'' + \left(\dfrac{\lambda^2}{x} + 1\right)w = 0, x > 0$;

   (iv) $w'' + \left(1 + \dfrac{1}{\lambda x}\right)w = 0, x > 0$;

   (v) $w'' + \lambda^2(1 - x^2)w = 0, x > 1$;

   (vi) $w'' + \lambda^2(1 - x^2)w = 0, x < 1$;

   (vii) $w'' + (\lambda + \tfrac{1}{2} - \tfrac{1}{4}x^2)w = 0$;

   (viii) $xw'' + (\lambda - x)w' - \lambda w = 0, x \neq 0$;

   (ix) $w'' + \left(\lambda^2 - \dfrac{\alpha}{x^2}\right)w = 0, \alpha \text{ real and constant}, |x| \neq 0$;

   (x) $w'' + \lambda^2 \sin^2 x \, w = 0, x > 0$.

2. Show that the asymptotic solutions of

   $$w'' + \{\lambda^2 \phi_0(x) + \lambda \phi_1(x) + \phi_2(x)\}w = 0 \text{ as } \lambda \to \infty,$$

are given by (6.53), by looking for them in the form

$$w(x; \lambda) \sim e^{\lambda W(x)} \, U(x) \left\{ 1 + \sum_{n=1}^{\infty} V_n(x) \lambda^{-n} \right\}.$$

where $W(x)$, $U(x)$, $V_n(x)$, $n = 1, 2, \ldots$, have to be determined. (This is the usual WKB-procedure.)

3. Find the uniformly valid asymptotic solutions as $\lambda \to \infty$ and for $x \leqq O(\lambda)$ to the following boundary-value problems:

(i) $\quad w'' + \left(\lambda + \dfrac{1}{x} \right) w = 0,$

(ii) $\quad w'' + \left(1 + \dfrac{1}{\lambda x} \right) w = 0$ $\quad \left. \right\} \; w(1) = a, \; w'(1) = b;$

(iii) $\quad w'' + \lambda x w = 0,$

(iv) $\quad w'' - \lambda x w = 0,$

(v) $\quad w'' + (\lambda^2 + x^2) w = 0, \quad w(0) = a, \quad w'(0) = b;$

(vi) $\quad w'' + \left(1 + \dfrac{\lambda}{x} \right) w = 0, \quad w(1) = a, \quad w'(1) = b.$

4. Find the first few terms in the uniformly valid asymptotic expansion for $t = O(1/\varepsilon)$ of the solution of the boundary-value problem

$$w'' + f(\varepsilon t; \varepsilon) w = 0, \quad w(0; \varepsilon) = a, \quad w'(0; \varepsilon) = b,$$

where

$$f(\varepsilon t; \varepsilon) \sim \Phi_0(\varepsilon t) + \varepsilon \Phi_1(\varepsilon t) + \varepsilon^2 \Phi_2(\varepsilon t) + \ldots \text{ as } \varepsilon \to 0$$

and $\Phi_n(\varepsilon t)$, $n = 0, 1, 2, \ldots$, are continuous twice-differentiable functions of εt, $\Phi_0(\varepsilon t) \neq 0$ for any εt.

5. Use the procedure in this section for finding the first few terms in the uniformly valid asymptotic expansion, as $\varepsilon \to 0$, of solutions of the following boundary-value problems for the range $0 \leqq x \leqq O(1/\varepsilon)$:

(i) $\quad w'' - \dfrac{1}{(1+\varepsilon x)^2} \, w = 0,$

(ii) $\quad w'' + (1 + \varepsilon x + \varepsilon^2 x^2) w = 0,$ $\quad \left. \right\} \; w(0) = a, \; w'(0) = b;$

(iii) $\quad \varepsilon w'' - (1 - \varepsilon x + \varepsilon^2 x^2) w = 0,$

(iv) $\quad w'' + (1 + (\varepsilon x)^{\frac{1}{2}}) w = 0.$

6. A certain medium is such that the speed of propagation $c(x)$ of small-amplitude high-frequency, denoted by ω, oscillations decreases as $1/x\omega$ with distance moved through the medium. If the governing equation for the wave disturbance is $\psi_{tt} = c^2(x) \psi_{xx}$ find an asymptotic wave form for $x > 0$ for high-frequency oscillations. Find the form if $c(x) = 1/(x+\omega)$.

7. For spherically symmetric situations under certain conditions, the Schrödinger equation is of the form

$$W''(r; \lambda) - \left\{ \lambda^2 - V(r) - \frac{n(n+1)}{r^2} \right\} W(r; \lambda) = 0,$$

where n is a positive integer. By first transforming from r to x, by writing $r = e^x$, remove the singularity of the equation at $r = 0$ to $x = -\infty$. Transforming the resultant equation by writing $W = e^{\frac{1}{2}x}w$, obtain an equation for $w(x; \lambda)$, and hence obtain asymptotic solutions to the spherically symmetric equation as $\lambda \to \infty$.

6.3. Transition points

We saw in §6.2 that the character of the asymptotic solution of (6.43) with (6.44) was crucially dependent on the sign of $\phi_0(x)$. Generally, points at which $f(x; \lambda)$ is zero are called *transition points*. In this section we discuss the simplest kind of transition point by considering the Liouville equation

$$w'' + \lambda^2 \, \phi_0(x)w = 0, \tag{6.118}$$

in which there is a single transition point, in the range of interest of x, which is a *simple zero* of $\phi_0(x)$. The concepts required in more complicated types of transition points are in essence the same as those discussed below. If the zero of $\phi_0(x)$ is at $x = x_c$, then by a trivial change of coordinate we can make it the origin and so, without loss of generality, we consider (6.118) where the origin is a simple zero, that is

$$\phi_0(0) = 0, \quad \phi_0'(0) \neq 0. \tag{6.119}$$

To be specific, let us consider $\phi_0'(0) > 0$ and so

$$\left. \begin{array}{l} \phi_0(x) < 0, \quad x < 0 \\ \phi_0(x) > 0, \quad x > 0 \\ \phi_0'(0) = v^2 > 0, \end{array} \right\} \tag{6.120}$$

where v^2 is some positive number defined by the given $\phi_0(x)$. For x *not* in the neighbourhood of $x = 0$, (6.55) gives one form of the two asymptotic solutions of (6.118) as $\lambda \to \infty$. For $x > 0$, $\phi_0(x) > 0$ and, using (6.55), we may write the general solution $w(x; \lambda)$, which is oscillatory in this case, in the form

$$w(x; \lambda) \sim \{\phi_0(x)\}^{-\frac{1}{4}} \{a_1 \exp(i\lambda \int^{x} \phi_0^{\frac{1}{2}} \, dx')$$

$$+ b_1 \exp(-i\lambda \int^{x} \phi_0^{\frac{1}{2}} \, dx')\},$$

$$x > 0, \text{ as } \lambda \to \infty, \tag{6.121}$$

or, in the trigonometric form which we shall use below,

$$w(x; \lambda) \sim \{\phi_0(x)\}^{-\frac{1}{4}} \{c_1 \cos (\lambda \int^{x} \phi_0^{\frac{1}{2}} \, dx') +$$

$$+ d_1 \sin (\lambda \int^{x} \phi_0^{\frac{1}{2}} \, dx')\},$$

$$x > 0, \text{ as } \lambda \to \infty, \tag{6.122}$$

where c_1 and d_1 (or a_1 and b_1) are arbitrary constants.

When $x < 0$, $\phi_0(x) < 0$ from (6.120), and the asymptotic solution of (6.118), using (6.55) again, may be written as

$$w(x; \lambda) \sim |\phi_0(x)|^{-\frac{1}{4}} \{A_1 \exp(\lambda \int^{x} |\phi_0|^{\frac{1}{2}} \, dx')$$

$$+ B_1 \exp(-\lambda \int^{x} |\phi_0|^{\frac{1}{2}} \, dx')\},$$

$$x < 0, \text{ as } \lambda \to \infty \tag{6.123}$$

for x *not* in the vicinity of $x = 0$. Here A_1 and B_1 are arbitrary constants. The solution (6.123) is exponential rather than oscillatory in character. There is a transition solution valid *near $x = 0$*. A typical solution for all x might look like Fig. 6.1.

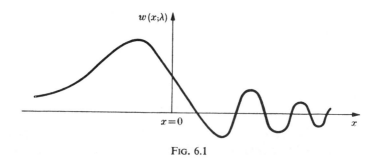

FIG. 6.1

There are two main problems associated with a transition point which we would like answered. The first is to find the relationship between c_1, d_1 (or a_1, b_1), and A_1, B_1, so that (6.122) (or (6.121)) and (6.123) are asymptotic forms of the *same* solution $w(x; \lambda)$ of (6.118). Boundary conditions may be given in either or both of the ranges $x > 0$ and $x < 0$. The second problem is to determine the form of the solution $w(x; \lambda)$ *in* the transition region which joins (6.122) on to (6.123) which, were they valid at $x = 0$, would be singular there. Since (6.122) and (6.123) are not valid solutions near $x = 0$, we must look for another form of the solution $w(x; \lambda)$ of (6.118) which does hold near $x = 0$, which is not singular there and which effects the join. This is a splendid example of an asymptotic matching process. We are interested, then, in the range of *small* x and so we may approximate $\phi_0(x)$ in this transition region by its linear approximation and write, using (6.120),

$$\phi_0(x) \doteqdot \phi_0'(0)x = v^2 x. \tag{6.124}$$

The approximate equation in the transition region, therefore, is, from (6.118) with (6.124),

$$w'' + \lambda^2 v^2 xw = 0. \tag{6.125}$$

This is a form of *Airy's equation* (3.41) discussed in §3.2 and in exercises 3 and 4 there; various forms of it were also considered in detail in §6.1. As mentioned in §3.2, it is also a special form of Bessel's equation and the solution of (6.125) may be written (see for example, Watson (1952)) in terms of Bessel functions of the first kind of order $\pm 1/3$, namely $x^{\frac{1}{2}}J_{\frac{1}{3}}(\frac{2}{3}\lambda vx^{\frac{3}{2}})$, $x^{\frac{1}{2}}J_{-\frac{1}{3}}(\frac{2}{3}\lambda vx^{\frac{3}{2}})$. The subsequent analysis may be carried through using such Bessel function solutions (exercise 2). However, since we have already studied, in some detail, Airy's equation and specifically its Airy function solutions in §3.2 (where for convenience we used a negative sign in the Airy equation (3.41) as compared with (6.125)), with exercises 3 and 4 there, we develop below the transition theory and solution using these results.

Solutions of (6.125) are the Airy functions $\mathrm{Ai}((\lambda v)^{\frac{2}{3}}x)$ and $\mathrm{Bi}((\lambda v)^{\frac{2}{3}}x)$ given as integrals by (3.47) and exercise 3 in §3.2, namely

$$\left.\begin{array}{l} \mathrm{Ai}\,((\lambda v)^{\frac{2}{3}}x) = \dfrac{1}{2\pi i}\int_{C_1} \exp((\lambda v)^{\frac{2}{3}}xz - \tfrac{1}{3}z^3)\mathrm{d}z, \\[3mm] \mathrm{Bi}\,((\lambda v)^{\frac{2}{3}}x) = \dfrac{1}{2\pi}(\int_{C_1} - \int_{C_3})\exp((\lambda v)^{\frac{2}{3}}xz - \tfrac{1}{3}z^3)\mathrm{d}z, \end{array}\right\} \tag{6.126}$$

where C_1, C_2, and C_3 are the contours in Fig. 3.4. The general solution of (6.125) is

$$w(x; \lambda) = a \text{ Ai} ((\lambda v)^{\frac{2}{3}}x) + b \text{ Bi} ((\lambda v)^{\frac{2}{3}}x), \qquad (6.127)$$

where a and b are arbitrary constants. We must now find the appropriate linear combination, that is a and b, in (6.127) which as the argument or simply $\lambda x \to -\infty$, that is $x < 0$, will join smoothly on to the asymptotic form of $w(x; \lambda)$ in (6.123) while for $\lambda x \to \infty$, that is $x > 0$, it will join smoothly on to the asymptotic form of $w(x; \lambda)$ in (6.122). To do this we need the asymptotic forms of the Airy functions $\text{Ai} ((\lambda v)^{\frac{2}{3}}x)$ and $\text{Bi} ((\lambda v)^{\frac{2}{3}}x)$ in (6.127) as $\lambda x \to \pm\infty$: we use the results of §3.2 on (6.126).

The functions $\text{Ai} ((\lambda v)^{\frac{2}{3}}x)$ and $\text{Bi} ((\lambda v)^{\frac{2}{3}}x)$ are integral functions (see §3.2) and so are not singular at the origin $x = 0$: a straightforward series solution approach also shows this. Thus $w(x; \lambda)$ in (6.126) is not singular at $x = 0$.

The asymptotic forms as $\lambda \to \infty$ of $\text{Ai} ((\lambda v)^{\frac{2}{3}}x)$ and $\text{Bi} ((\lambda v)^{\frac{2}{3}}x)$ obtained by the method of steepest descents in §3.2[†] are, for $x > 0$ (§3.2 exercise 4 since the Airy equation (3.41) is written in the form $d^2w/d\lambda^2 - \lambda w = 0$ and so $\lambda \gtrless 0$ there corresponds here, with (6.125) respectively to $x \lessgtr 0$).

$$\left.\begin{array}{l} \text{Ai} ((\lambda v)^{\frac{2}{3}}x) \sim \dfrac{1}{\sqrt{\pi} (\lambda v)^{\frac{1}{6}} x^{\frac{1}{4}}} \sin\left(\tfrac{2}{3}\lambda v\, x^{\frac{3}{2}} + \tfrac{\pi}{4}\right) \\[3mm] \text{Bi} ((\lambda v)^{\frac{2}{3}}x) \sim \dfrac{1}{\sqrt{\pi} (\lambda v)^{\frac{1}{6}} x^{\frac{1}{4}}} \cos\left(\tfrac{2}{3}\lambda v\, x^{\frac{3}{2}} + \tfrac{\pi}{4}\right) \end{array}\right\} \begin{array}{l} x > 0, \text{ as } \lambda x \to \infty, \\[4mm] (6.128) \end{array}$$

and, for $x < 0$ (§3.2, equation (3.56) and exercise 3),

$$\left.\begin{array}{l} \text{Ai} ((\lambda v)^{\frac{2}{3}}x) \sim \dfrac{1}{2\sqrt{\pi} (\lambda v)^{\frac{1}{6}} |x|^{\frac{1}{4}}} e^{-\frac{2}{3}\lambda v |x|^{\frac{3}{2}}} \\[4mm] \text{Bi} ((\lambda v)^{\frac{2}{3}}x) \sim \dfrac{1}{\sqrt{\pi} (\lambda v)^{\frac{1}{6}} |x|^{\frac{1}{4}}} e^{\frac{2}{3}\lambda v |x|^{\frac{3}{2}}} \end{array}\right\} \begin{array}{l} x < 0, \text{ as } \lambda x \to -\infty. \\[4mm] (6.129) \end{array}$$

Using (6.128) and (6.129), we must now compare $w(x; \lambda)$ in (6.127) with $\lambda|x|$ large, with the asymptotic forms (6.122) and (6.123). In these latter two equations, however, $\phi_0(x)$ appears and not v^2x as in (6.128) and (6.129). Nevertheless we may use v^2x for $\phi_0(x)$ in

† Alternatively, see Jeffreys (1962).

(6.122) and (6.123) for comparison with the transition solution (6.127) with $\lambda|x| \to \infty$ by simply choosing x sufficiently small, but not zero, so that $\phi_0(x) \doteqdot v^2 x$, and λ sufficiently large so that (6.128) and (6.129) *are* valid. This is sufficient to determine the relationships between the constants c_1, d_1, in (6.122), A_1, B_1, in (6.123) and a, b, in (6.127). For this purpose (6.122) becomes, with $\phi_0(x) \doteqdot v^2 x$, and first taking $x > 0$,

$$w(x; \lambda) \sim \frac{1}{v^{\frac{1}{4}} x^{\frac{1}{4}}} \{c_1 \cos (\lambda \int vx^{\frac{1}{2}} \, dx) + d_1 \sin (\lambda \int vx^{\frac{1}{2}} \, dx)\},$$

$$x > 0, \text{ as } \lambda \to \infty. \tag{6.130}$$

We now compare this with (6.129) on substituting for $\text{Ai}\,((\lambda v)^{\frac{2}{3}} x)$ and $\text{Bi}\,((\lambda v)^{\frac{2}{3}} x)$ from (6.128), namely

$$w(x; \lambda) \sim \frac{1}{\sqrt{\pi}\,(\lambda v)^{\frac{1}{6}} x^{\frac{1}{4}}} \left\{ a \sin \left(\tfrac{2}{3}\lambda v \, x^{\frac{3}{2}} + \frac{\pi}{4} \right) + \right.$$

$$\left. + b \cos \left(\tfrac{2}{3}\lambda v \, x^{\frac{3}{2}} + \frac{\pi}{4} \right) \right\},$$

$$x > 0, \text{ as } \lambda x \to \infty. \tag{6.131}$$

If we choose $\lambda \int v \, x^{\frac{1}{2}} \, dx$ as

$$\lambda \int_0^x v \, x'^{\frac{1}{2}} \, dx' = \tfrac{2}{3}\lambda v \, x^{\frac{3}{2}}, \tag{6.132}$$

(6.130) becomes

$$w(x; \lambda) \sim \frac{1}{v^{\frac{1}{4}} x^{\frac{1}{4}}} (c_1 \cos \tfrac{2}{3}\lambda v \, x^{\frac{3}{2}} + d_1 \sin \tfrac{2}{3}\lambda v \, x^{\frac{3}{2}}),$$

$$x > 0, \text{ as } \lambda x \to \infty, \tag{6.133}$$

which, on comparison with (6.131), on expanding the sine and cosine, immediately gives

$$c_1 = \frac{v^{\frac{1}{3}}}{\sqrt{(2\pi)}\,\lambda^{\frac{1}{6}}} (a+b), \quad d_1 = \frac{v^{\frac{1}{3}}}{\sqrt{(2\pi)}\,\lambda^{\frac{1}{6}}} (a-b), \tag{6.134}$$

which determine a and b in terms of c_1 and d_1 (and hence a_1 and b_1 of (6.121)).

In an analogous way, for $x < 0$, we get for (6.123), on using $\phi_0(x) \doteqdot v^2 x$ and (6.132) as before,

$$w(x; \lambda) \sim \frac{1}{v^{\frac{1}{2}} |x|^{\frac{1}{4}}} (A_1 \, e^{\frac{2}{3}\lambda v|x|^{\frac{3}{2}}} + B_1 \, e^{-\frac{2}{3}\lambda v|x|^{\frac{3}{2}}}),$$

$$x < 0, \text{ as } \lambda x \to -\infty, \tag{6.135}$$

which must now be compared with (6.127) on using (6.129) for the asymptotic forms of Ai $((\lambda v)^{\frac{2}{3}} x)$ and Bi $((\lambda v)^{\frac{2}{3}} x)$, namely

$$w(x; \lambda) \sim \frac{1}{\sqrt{\pi} \, (\lambda v)^{\frac{1}{6}} |x|^{\frac{1}{4}}} (2a \, e^{-\frac{2}{3}\lambda v|x|^{\frac{3}{2}}} + b \, e^{\frac{2}{3}\lambda v|x|^{\frac{3}{2}}}),$$

$$x < 0, \text{ as } \lambda x \to -\infty. \tag{6.136}$$

Comparison of (6.135) and (6.136) immediately gives

$$A_1 = \frac{b \, v^{\frac{1}{2}}}{\sqrt{\pi} \, \lambda^{\frac{1}{6}}}, \quad B_1 = \frac{2a \, v^{\frac{1}{2}}}{\sqrt{\pi} \, \lambda^{\frac{1}{6}}}, \tag{6.137}$$

which give A_1 and B_1 in terms of a and b or, of course, a and b in terms of A_1 and B_1. Relations (6.134) and (6.137) now give the relation between c_1 and d_1 (or a_1 and b_1) and A_1 and B_1 as

$$c_1 = \frac{1}{\sqrt{2}} (\tfrac{1}{2} B_1 + A_1), \quad d_1 = \frac{1}{\sqrt{2}} (\tfrac{1}{2} B_1 - A_1). \tag{6.138}$$

Relations (6.138) alternatively may be solved to give A_1 and B_1 in terms of c_1 and d_1.

We have thus achieved what we set out to do, namely, to relate the solution on one side of the transition point (6.122) to the appropriate one (6.123) on the other side by obtaining two relations, namely (6.138), between the four constants c_1, d_1, and A_1, B_1. In the process we have also determined the approximate asymptotic transition solution which holds in the vicinity of $x = 0$ since a and b in (6.127) are also determined from (6.137) with (6.138). It is clear that it is immaterial which two constants of c_1, d_1 (or a_1, b_1), A_1, B_1, are given by the boundary conditions.

This is an illustration of asymptotic matching. Here the solution in the outer region, namely (6.122) and (6.123), for small x is made to join on to that valid in the inner region, namely (6.127), as its independent variable, here $(\lambda v)^{\frac{2}{3}} x$, becomes large. This matching process, of which this is a good example, is discussed in detail by Van Dyke (1964) and Cole (1966).

The case $\phi_0'(0) < 0$ (compare with (6.120)) can clearly be treated in exactly the same way.

More complicated transition points, for example, when $\phi_0(x)$ has other than simple zeros or when it has zeros which are close together, are possible. The study of these can become quite involved but the essential concept, however, is similar to the above. The analysis involves, also as above, matching asymptotic expansions to obtain a uniformly valid asymptotic solution.

Exercises

1. Consider the equation
$$w'' + \lambda^2(x^2 - 1)w = 0, \quad 0 \le x \le 2,$$
   as $\lambda \to \infty$ with $w(0) = 0$, $w'(0) = 1$. First obtain asymptotic forms for $w(x; \lambda)$ as $\lambda \to \infty$ for x not near the transition point $x = 1$ using the method from §6.2 (exercises 1 (v) and 1 (vi)). Then find the approximate transition solution and hence obtain a uniformly valid asymptotic solution for all $0 \le x \le 2$, carefully noting the connections between the three parts of the solution.

2. Given that $x^{\frac{1}{2}} J_{\pm\frac{1}{3}} (\frac{1}{3}\lambda vx^{\frac{3}{2}})$ are the Bessel function solutions of (6.125), obtain the relationships between c_1 and d_1 in (6.122) and A_1 and B_1 in (6.123) using the following asymptotic expansions of $x^{\frac{1}{2}} J_{\pm\frac{1}{3}} (\frac{1}{3}\lambda vx^{\frac{3}{2}})$ as $\lambda x \to \pm\infty$:

$$\left.\begin{array}{l} x^{\frac{1}{2}} J_{\frac{1}{3}} (\tfrac{1}{3}\lambda vx^{\frac{3}{2}}) \sim \left(\dfrac{3}{\lambda v\pi}\right)^{\frac{1}{2}} x^{-\frac{1}{4}} \cos\left(\tfrac{1}{3}\lambda vx^{\frac{3}{2}} - \dfrac{5}{12}\pi\right) \\[4mm] x^{\frac{1}{2}} J_{-\frac{1}{3}} (\tfrac{1}{3}\lambda vx^{\frac{3}{2}}) \sim \left(\dfrac{3}{\lambda v\pi}\right)^{\frac{1}{2}} x^{-\frac{1}{4}} \cos\left(\tfrac{1}{3}\lambda vx^{\frac{3}{2}} - \dfrac{1}{12}\pi\right) \end{array}\right\} \begin{array}{l} x > 0, \\ \lambda x \to \infty; \end{array}$$

$$\left.\begin{array}{l} x^{\frac{1}{2}} J_{\frac{1}{3}} (\tfrac{1}{3}\lambda vx^{\frac{3}{2}}) \sim \dfrac{1}{2} \left(\dfrac{3}{\lambda v\pi}\right)^{\frac{1}{2}} |x|^{-\frac{1}{4}} \left(\exp\left(-\tfrac{2}{3}\lambda v |x|^{\frac{3}{2}} - \dfrac{i\pi}{6}\right) - \right. \\[3mm] \hspace{3cm} \left. - \exp\left(\tfrac{2}{3}\lambda v |x|^{\frac{3}{2}}\right)\right), \\[4mm] x^{\frac{1}{2}} J_{-\frac{1}{3}} (\tfrac{1}{3}\lambda vx^{\frac{3}{2}}) \sim \dfrac{1}{2} \left(\dfrac{3}{\lambda v\pi}\right)^{\frac{1}{2}} |x|^{-\frac{1}{4}} \left(\exp\left(-\tfrac{2}{3}\lambda v |x|^{\frac{3}{2}} + \dfrac{i\pi}{6}\right) + \right. \\[3mm] \hspace{3cm} \left. + \exp\left(\tfrac{2}{3}\lambda v |x|^{\frac{3}{2}}\right)\right). \end{array}\right\} \begin{array}{l} x < 0, \\ \lambda x \to -\infty. \end{array}$$

7

Singular perturbation methods

7.1. Basic concepts and introduction to the method of matched expansions

IN §6.2 in the last chapter we briefly discussed, by way of example, some singular perturbation problems in linear differential equations which could be solved using the exponential method developed there. In this chapter we discuss some more generally applicable singular perturbation techniques. These are extremely powerful and let us consider *nonlinear* problems, involving a small or large parameter, with a view to extracting the principle features of the solutions which cannot be found by classical methods. A fundamental property of nonlinear problems that we shall be interested in is that the main features of the solutions are not contained in the linearized problem as we see in §7.2 below.

We first give, in this section, a general definition of what a singular perturbation problem is and then discuss intuitively, by way of examples, the method of matched expansions. In §7.2 the technique is also described by way of worked problems. Singular perturbation theory is now a large and important subject. There are several good books which specifically deal with it and contain many detailed case studies, for example, Cole (1968), Kevorkian and Cole (1981), Nayfeh (1973), O'Malley (1974) and a succinct but very readable article by Cohen (1977). There is a fuller discussion of matched expansion methods and a variety of examples which come from the bio-medical sciences in the book by Murray (1977) on differential equation models in biology.

Let ε be a small real (positive or negative) parameter, that is, $|\varepsilon| \ll 1$, and the general mathematical problem be denoted by

$$L(u, r, t; \varepsilon) = 0, \tag{7.1}$$

with boundary conditions

$$B(u, r, t; \varepsilon) = 0, \tag{7.2}$$

where $u(r, t; \varepsilon)$ is the dependent variable, r and t the independent variables, and L and B operations which may, for example, be derivatives or integrals involving u. If the solution $u(r, t; \varepsilon)$ depends *analytically* on ε in some domain $0 \leq |\varepsilon| < \varepsilon_0$ for some positive value ε_0 and r and t in the domain of interest then as $\varepsilon \to 0$ the solution can be written as a simple Taylor series in ε, namely,

$$u(r, t; \varepsilon) = u_0(r, t) + \varepsilon u_1(r, t) + \ldots, \quad \varepsilon \to 0, \tag{7.3}$$

where $u_0, u_1, \ldots$ can be determined systematically by expanding (7.1) and (7.2) as series expansions in ε and equating powers of ε. Here $u_0(r, t) = \lim_{\varepsilon \to 0} u(r, t; \varepsilon)$ satisfies

$$L(u_0, r, t, 0) = 0, \quad B(u_0, r, t; 0) = 0. \tag{7.4}$$

The solution (7.3) is simply a *regular* or *nonsingular* perturbation solution. Mathematically (and in any practical situation which gives rise to such a problem) such a solution is never very interesting since if $u_0(r, t)$ is known, the more accurate approximation $u_0 + \varepsilon u_1$ is simply a small $O(\varepsilon)$ perturbation from u_0. On the other hand if $\lim_{\varepsilon \to 0} u(r, t; \varepsilon)$ is not obtained uniformly, that is $u(r, t; \varepsilon)$ is *not* analytic as $\varepsilon \to 0$, an expansion such as (7.3) is *not* uniformly valid: the problem (7.1)–(7.2) is then a *singular perturbation* one. It should be noted at this stage that the specification of the domain in r, t and ε is often crucial in discussing asymptotic expansions. The examples below will illustrate this point.

To illustrate and motivate the method of matched asymptotic expansions consider the simple scalar equation for $u(x; \varepsilon)$ for $0 \leq x \leq 1$, namely

$$\varepsilon u'' + u' = 0, \quad |\varepsilon| \ll 1, \tag{7.5}$$

where the prime denotes differentiation with respect to x and the boundary conditions are as yet unspecified. Here (7.5) corresponds to $L(u; \varepsilon) = 0$. The exact solution of (7.5) is

$$u(x; \varepsilon) = a_1(\varepsilon) e^{-x/\varepsilon} + a_2(\varepsilon), \tag{7.6}$$

where a_1 and a_2 are constants of integration which may depend on ε. Now consider two different sets of boundary conditions, one associated with a two-point boundary value problem and another with an initial value problem. We first discuss the former and take

$$u(0; \varepsilon) = U_0, \quad u(1; \varepsilon) = U_1, \tag{7.7}$$

where U_0 and U_1 ($\neq U_0$) are, for simplicity only, independent of ε. The solution (7.6) satisfying (7.7) is

$$u(x; \varepsilon) = \frac{U_0[e^{(1-x)/\varepsilon} - 1]}{[e^{1/\varepsilon} - 1]} + \frac{U_1[1 - e^{-x/\varepsilon}]}{[1 - e^{-1/\varepsilon}]}. \tag{7.8}$$

The function is not analytic in ε as $\varepsilon \to 0$; $u(x; \varepsilon)$ clearly does not tend to a definite limit as $\varepsilon \to 0$ unless we bound x away from $x = 0$ and $x = 1$. In what follows we must distinguish the two cases $\varepsilon \to 0$, ε positive and $\varepsilon \to 0$, ε negative; these cases are conventionally written as $\varepsilon \downarrow 0$ and $\varepsilon \uparrow 0$ respectively. From (7.8) we have as $\varepsilon \downarrow 0$

$$\left. \begin{aligned} u(x; \varepsilon) &\sim U_0 e^{-x/\varepsilon} + U_1, \quad 0 < x \leq 1, \quad 0 < \varepsilon \ll 1, \\ &\to U_1, \quad 0 < x \leq 1 \text{ as } \varepsilon \downarrow 0. \end{aligned} \right\} \tag{7.9}$$

For a negative $\varepsilon \uparrow 0$,

$$\left. \begin{aligned} u(x; \varepsilon) &\sim U_0 + U_1 e^{(1-x)/\varepsilon}, \quad 0 \leq x < 1, \quad 0 < |\varepsilon| \ll 1, \\ &\to U_0, \quad 0 \leq x < 1 \text{ as } \varepsilon \uparrow 0. \end{aligned} \right\} \tag{7.10}$$

Continuing with the ε-positive case, the limiting solution U_1 in (7.8) as $\varepsilon \downarrow 0$ satisfies the boundary condition at $x = 1$ in (7.7) but *not* at $x = 0$. If we now let $\varepsilon \downarrow 0$ in the differential equation (7.5) we get

$$u' = 0 \Rightarrow u(x) = \text{constant}.$$

With the constant of integration taken as U_1 this solution satisfies the second boundary condition (7.7) but not the first. On the other hand with the constant taken to be U_0 we can satisfy the first but not the second of (7.7).

In this example and those of a matched asymptotic type we would expect difficulties in satisfying the boundary conditions, since on letting $\varepsilon \to 0$ in (7.5) the order of the differential equation is reduced and so in general we lose a constant of integration which means we can satisfy one fewer boundary condition. The fact that ε multiplies the *highest* derivative in a differential equation (partial and ordinary) is an immediate indication that the problem is, except in irregular circumstances, a singular perturbation one and usually of the matched asymptotic kind. There are other classes of singular perturbation problems in which a small parameter does *not* multiply the highest derivative in the equation and some of the major ones we consider below.

Still considering $\varepsilon > 0$ the solution (7.8) *near* $x = 0$ is

$$u(x; \varepsilon) \sim U_0\, e^{-x/\varepsilon} + U_1(1 - e^{-x/\varepsilon}) \text{ as } \varepsilon \downarrow 0, \qquad (7.11)$$

which shows that

$$\lim_{x\downarrow 0} \lim_{\varepsilon\downarrow 0} u(x;\varepsilon) = U_1 \neq U_0 = \lim_{\varepsilon\downarrow 0} \lim_{x\downarrow 0} u(x;\varepsilon).$$

That is the limits $\varepsilon \downarrow 0$ and $x \downarrow 0$ are *not* commutative: it is clearly important to know which process we want.

Note, from (7.11), that there is a *singular region* or boundary layer near $x = 0$, that is a small $O(\varepsilon)$ domain in x in which the solution changes very rapidly, here from the boundary value U_0 to the limiting value

$$\lim_{\substack{\varepsilon\downarrow 0 \\ x\neq 0}} u(x;\varepsilon) = U_1.$$

The gradient of u at $x = 0$ is $O(1/\varepsilon)$; in fact from (7.8) (and from (7.11)),

$$u'(0;\varepsilon) = \frac{1}{\varepsilon}(U_1 - U_0).$$

Figure 7.1(a) schematically illustrates the solution and the singular region when $U_1 > U_0$. On the other hand when negative $\varepsilon \uparrow 0$ the singular region is at $x = 1$ as in Fig. 7.1(b): this can be seen from a parallel analysis to the $\varepsilon \downarrow 0$, $x \downarrow 0$ case but now with $\varepsilon \uparrow 0$, $x \uparrow 1$.

Suppose we now consider (7.5) for small ε and try to derive the asymptotic solution $u(x;\varepsilon)$ by a method other than using the exact solution (as we certainly have to do in problems where the exact solution cannot be obtained analytically either easily or at all). The numerical solution of singular perturbation problems can present considerable difficulties: a preliminary singular perturbation analysis is often a necessary prerequisite to writing a suitable numerical programme.

Since ε is small suppose we naively look for a regular (Taylor series in ε) perturbation solution of the form (7.3) namely

$$u(x;\varepsilon) \sim u_0(x) + \varepsilon u_1(x) + \ldots, \text{ as } \varepsilon \to 0. \qquad (7.12)$$

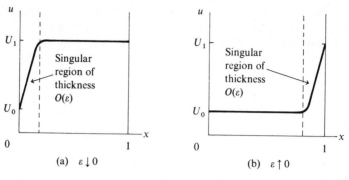

FIG. 7.1

Substituting this into (7.5) and systematically equating the coefficients of powers of ε we get

$$u'_0 = 0, \quad u'_n = -u''_{n-1}, \quad n \geq 1 \Rightarrow u'_n = 0, \quad n \geq 0,$$

which have constants as solutions, say $u_n = a_n, n \geq 0$. The solution (7.12) is then simply a constant,

$$u(x; \varepsilon) \sim a_0 + \varepsilon a_1 + \dots, \tag{7.13}$$

which clearly cannot satisfy both boundary conditions (7.7). Thus a solution of the form (7.12) is *not* a uniformly valid asymptotic solution for small ε for all x in the domain $0 \leq x \leq 1$. However, we can make (7.13) satisfy one of the boundary conditions (7.7) by choosing $a_{n \geq 1} = 0$ and $a_0 = U_0$ or $a_0 = U_1$. Since such a solution satisfies one boundary condition we expect it to be a valid approximation solution for some subdomain of x in $0 \leq x \leq 1$. If this were not the case it would mean that $\varepsilon u''$ would have to be of the same order as u' everywhere since on neglecting $\varepsilon u''$ the resulting solution was nowhere valid.

The underlying assumption in (7.12) is that $u_{n \geq 0}(x)$ and their first two derivatives are continuous functions of x and are all $O(1)$. But, as we saw above such an assumption did not give a uniformly valid asymptotic solution. It could only satisfy one boundary condition. Thus we must introduce a transformation in x which involves ε in such a way that as $\varepsilon \to 0$, $\varepsilon u''$ contributes to the $O(1)$ solution in the x-domain where the solution (7.13) is not valid, that is near one or other boundary. (Note that since the problem is *linear* in u a transformation of u could not help so it has to be one involving x.)

Suppose in the first instance we choose the regular or non-singular perturbation solution (7.13) to satisfy the boundary condition at $x = 1$ that is we choose $a_0 = U_1$, $a_{n \geq 1} = 0$. The resulting solution $u(x; \varepsilon) \sim U_1$ does not satisfy the boundary condition at $x = 0$ so we transform x by introducing

$$\xi = x/\varepsilon^\alpha,$$

where α must be determined although we anticipate $\alpha > 0$. With $\alpha > 0$ this transformation stretches out the immediate neighbourhood of $x = 0$ since, for any positive x, ξ is large because ε is small. In fact for $x > 0$, however small, $\xi \to \infty$ as $\varepsilon \downarrow 0$.

Writing $u(x; \varepsilon) = \bar{u}(\xi; \varepsilon)$, equation (7.5) becomes

$$\varepsilon^{1-2\alpha}\bar{u}_{\xi\xi} + \varepsilon^{-\alpha}\bar{u}_\xi = 0 \Rightarrow \varepsilon^{1-\alpha}\bar{u}_{\xi\xi} + \bar{u}_\xi = 0.$$

If $\alpha = 1$ then as $\varepsilon \downarrow 0$ *with ξ fixed* the second-order term remains in the first approximation. The purpose of this 'stretching' transformtion is to retain the highest derivative under the limiting process and it is this which determines α: see exercise 2(ii) for an example where $\alpha \neq 1$. With this particularly simple equation, with $\alpha = 1$, ε does not appear in the transformed equation (this is exceptional). This equation governs the solution in the immediate neighbourhood of $x = 0$ which corresponds to the *complete* ξ-domain $0 \leq \xi < \infty$ since for $x = 0$, $\xi = 0$ and for $x > 0$, however small, $\xi = x/\varepsilon \to \infty$ as $\varepsilon \downarrow 0$. This equation is the same as the original one (7.5) *but* the boundary conditions will be different. We now require a solution to the last equation, with $\alpha = 1$, which satisfies the boundary condition $U_0 = u(0; \varepsilon) = \bar{u}(0; \varepsilon)$ and which must join onto $u(x; \varepsilon) = U_1$ as we leave the neighborhood of $x = 0$ which, as we saw above, is equivalent to $\xi \to \infty$. Such a solution is

$$\bar{u}(\xi; \varepsilon) = U_1 + (U_0 - U_1)e^{-\xi}.$$

In terms of the original variable x this gives the $O(1)$ asymptotic solution uniformly valid for all $0 \leq x \leq 1$ as

$$u(x; \varepsilon) \sim U_1 + (U_0 - U_1)e^{-x/\varepsilon} \text{ as } \varepsilon \downarrow 0, \quad (7.14)$$

that is the same as (7.11) obtained from the exact solution as $\varepsilon \downarrow 0$.

Suppose now we make (7.13) satisfy the boundary condition at $x = 0$ rather than at $x = 1$: that is we choose $a_0 = U_0$, $a_{n \geq 1} = 0$.

In this case the regular perturbation solution (7.13) is not uniformly valid at $x = 1$. If we follow the above procedure we now have to find a singular solution near $x = 1$ which will join the solution $u = U_0$ onto the boundary value $u(1; \varepsilon) = U_1$. With the above experience we introduce a new independent variable $\eta = (1-x)/\varepsilon$ and the differential equation (7.5) becomes, for $u(x; \varepsilon) = \tilde{u}(\eta; \varepsilon)$ say,

$$\tilde{u}_{\eta\eta} - \tilde{u}_{\eta} = 0.$$

In this case we require a solution $\tilde{u}(\eta; \varepsilon)$ which satisfies $\tilde{u}(0; \varepsilon) = U_1$ (here $x = 1$ corresponds to $\eta = 0$) and tends to the solution $u = U_0$ as we leave the immediate neighbourhood of $x = 1$, that is as $n \to \infty$. Here, however, since $\tilde{u}(\eta; \varepsilon) = \tilde{a} + \tilde{b}\, e^{\eta}$ with $\tilde{a}$, $\tilde{b}$ constants there is no such solution of the last equation and so there can be no singular domain near $x = 1$ as $\varepsilon \downarrow 0$. Thus in this case, namely $\varepsilon > 0$, the only uniformly valid asymptotic solution as $\varepsilon \downarrow 0$ is (7.14): that is there is a singular region, near $x = 0$ as in Fig. 7.1(a). Although strictly we must consider all possibilities, experience quickly indicates where the singular region is.

This example shows that even though the reduced $\varepsilon = 0$ form of (7.5) does not have a unique solution the singular perturbation procedure produces the correct asymptotic form of the exact solution as $\varepsilon \downarrow 0$. It can be shown in a similar way that if $\varepsilon < 0$ and $\varepsilon \uparrow 0$ the correct unique asymptotic form of (7.8) is again obtained. Here there is a singular region at $x = 1$ instead of at $x = 0$: this case is illustrated in Fig. 7.1(b).

Consider now the boundary conditions for (7.5) which correspond to an initial value problem, namely

$$u(0; \varepsilon) = U_0, \quad u'(0; \varepsilon) = V_0, \tag{7.15}$$

where U_0 and V_0 are independent of ε. Once again the nonsingular regular perturbation solution is (7.13) which again cannot satisfy both of (7.15). In this case however there is no choice for a_0 other than $a_0 = U_0$. The singular solution $\bar{u}(\xi; \varepsilon)$, with $\xi = x/\varepsilon$ as before, must again satisfy

$$\bar{u}_{\xi\xi} + \bar{u}_{\xi} = 0 \Rightarrow \bar{u}(\xi, \varepsilon) = \bar{a} + \bar{b}\, e^{-\xi}, \tag{7.16}$$

where a and b are constants to be determined. Now $\bar{u}(\xi; \varepsilon)$ must tend, as $\xi \to \infty$, to the regular solution u which here is $u = U_0$:

this gives $\bar{a} = U_0$. Also $\bar{u}(\xi; \varepsilon)$ must satisfy the second of (7.15) which gives, using (7.16)

$$V_0 = \frac{du}{dx}\bigg|_{x=0} = \frac{1}{\varepsilon}\frac{d\bar{u}}{d\xi}\bigg|_{\xi=0} \Rightarrow \bar{b} = -\varepsilon V_0$$

and so

$$\bar{u}(\xi; \varepsilon) = U_0 - \varepsilon V_0\, e^{-\xi}.$$

Thus the uniformly valid asymptotic solution for all $0 \leq x \leq 1$ which satisfies both conditions (7.15) is from the last equation

$$u(x; \varepsilon) \sim U_0 - \varepsilon V_0\, e^{-x/\varepsilon} \text{ as } \varepsilon \downarrow 0. \tag{7.17}$$

However the exact solution of (7.5) with boundary conditions (7.15) is

$$\left.\begin{aligned} u(x; \varepsilon) &= U_0 + \varepsilon V_0(1 - e^{-x/\varepsilon}) \\ &= (U_0 - \varepsilon V_0\, e^{-x/\varepsilon}) + O(\varepsilon) \text{ as } \varepsilon \downarrow 0. \end{aligned}\right\} \tag{7.18}$$

In this case, however, the asymptotic solution (7.17) differs from the exact solution by a term $O(\varepsilon)$ namely εV_0.

In this example with initial value boundary conditions the question now arises as to how the singular perturbation procedure determines the $O(\varepsilon)$ correction to (7.18) which we know from the exact solution to be εV_0. Such a procedure, namely the one known as the method of matched asymptotic expansions, will be developed in the next example; we also introduce an important concept of singular perturbation analysis.

From a verbal descriptive point of view the nonsingular part of the solution is usually referred to as the *outer* (or *nonsingular*) solution while the singular part, which in the above is where $\varepsilon u''$ is *not* $O(\varepsilon)$, is called the *inner* (or *singular*) solution. In the above example (7.13) is the outer solution while (7.14) and (7.17) are the inner solutions.

Consider the ordinary differential equation for $u(x; \varepsilon)$ for $0 < x < 1, 0 < \varepsilon \ll 1$, given by

$$\varepsilon u'' + u' + u = 0, \tag{7.19}$$

with boundary conditions

$$u(0; \varepsilon) = U_0, \quad u(1; \varepsilon) = U_1, \tag{7.20}$$

where U_0 and U_1 are constants independent of ε: (7.19) and (7.20) correspond respectively to (7.1) and (7.2). We want to find a uniformly valid asymptotic solution for ε small to any order in ε: we shall write $\varepsilon \to 0$ for $\varepsilon \downarrow 0$ in the following since $\varepsilon > 0$. Of course as before we can solve this (linear) equation exactly but we proceed as if we could not.

Away from the boundaries we expect the outer solution to be valid and we look for it in terms of a regular Taylor series in ε such as (7.12). Substituting (7.12) into (7.19) and systematically equating powers of ε we get

$$u_0' + u_0 = 0, \quad u_n' + u_n = -u_{n-1}'', \quad n \geq 1,$$

the solutions of which are

$$u_0(x) = a_0 e^{-x}, \quad u_1(x) = a_1 e^{-x} - a_0 x e^{-x}, \ldots . \qquad (7.21)$$

As we now expect, this regular perturbation solution cannot satisfy both boundary conditions and hence there is at least one singular region near $x = 0$ or $x = 1$. The following asymptotic procedure determines which it is just as in the example above. (The singular layer is near $x = 0$ in fact.)

Let us require (7.21) to satisfy the second of (7.20) in the first instance, that is

$$u_0(1) + \varepsilon u_1(1) + \ldots = U_1 \Rightarrow u_0(1) = U_1, \quad u_{n \geq 1}(1) = 0.$$

From (7.21) this determines the $a_{n \geq 0}$ as

$$a_0 = U_1 e, \quad a_1 = U_1 e, \ldots,$$

and so the outer (non-singular) solution valid in $0 < x \leq 1$ is

$$u(x; \varepsilon) = U_1 e^{(1-x)} + \varepsilon U_1 (1-x) e^{1-x} + O(\varepsilon^2) \qquad (7.22)$$

and it satisfies $u(1; \varepsilon) = U_1$. It is not a uniformly valid asymptotic solution, as $\varepsilon \to 0$, of (7.19) and (7.20) since it does not satisfy the boundary condition at $x = 0$ as is trivially checked.

We now look for the inner (singular) solution of (7.19) valid at $x = 0$ and in its neighbourhood. With the above worked example in mind we introduce the stretched variable $\xi = x/\varepsilon$ and look for the singular solution in the form

$$u(x; \varepsilon) = \bar{u}(\xi; \varepsilon) \sim \bar{u}_0(\xi) + \varepsilon \bar{u}_1(\xi) + \ldots, \quad \xi = x/\varepsilon. \qquad (7.23)$$

With this form the differential equation (7.19) becomes

$$\bar{u}_{\xi\xi} + \bar{u}_{\xi} + \varepsilon\bar{u} = 0$$

and equating powers of ε gives the sequence of recursive equations

$$\bar{u}_{0\xi\xi} + \bar{u}_{0\xi} = 0, \quad \bar{u}_{n\xi\xi} + \bar{u}_{n\xi} = -\bar{u}_{n-1}, \quad n \geq 1.$$

The solutions of these are

$$\left.\begin{array}{l} \bar{u}_0(\xi) = \bar{a}_0 + \bar{b}_0\, e^{-\xi}, \\ \bar{u}_1(\xi) = \bar{a}_1 + \bar{b}_1\, e^{-\xi} + \bar{a}_0(1-\xi) + \bar{b}_0\xi\, e^{-\xi}. \end{array}\right\} \tag{7.24}$$

We now require (7.23) to satisfy the first of the boundary conditions (7.20), that is at $\xi = 0$,

$$\bar{u}_0(0) + \varepsilon\bar{u}_1(0) + \ldots = U_0,$$

which, from (7.24), gives

$$\bar{a}_0 + \bar{b}_0 = U_0, \quad \bar{a}_1 + \bar{b}_1 + \bar{a}_0 = 0, \ldots. \tag{7.25}$$

These do *not* determine $\bar{a}_{n\geq 0}$ and $\bar{b}_{n\geq 0}$: we require another condition on each of the $\bar{u}_n(\xi)$.

The fact that we cannot determine the inner solution completely at this stage is the norm rather than the exception in singular perturbation problems. In fact usually neither the inner nor the outer solutions can be determined explicitly. What we require is that the inner solution must merge into the outer solution. This is the main point of the *matching* process in singular perturbation problems. The question now is how to effect such a match and in so doing to determine the unknown constants of integration which, in the above, are $\bar{a}_{n\geq 0}$ and $\bar{b}_{n\geq 0}$. The general problem of matching is nontrivial: here we give the procedure that works in most cases. The reader is referred to the books cited for a fuller and more sophisticated treatment.

In the previous example we simply matched the inner solution onto the outer solution and to $O(1)$ the complete uniformly valid solution was obtained immediately. Basic to the method of matched asymptotic expansions is the concept that the inner solution and outer solution *overlap* in some *intermediate* region. The existence of such a domain is not always easy to prove. In our examples the appropriate variable in the outer expansion is x while for the inner expansion it is x/ε near $x = 0$. An intermediate variable near the $x = 0$ boundary, say, may be written as

$$\left.\begin{array}{l} \zeta = \dfrac{x}{\alpha(\varepsilon)}, \\[2mm] \alpha(\varepsilon) \to 0, \quad \dfrac{\alpha(\varepsilon)}{\varepsilon} \to \infty, \text{ as } \varepsilon \to 0, \end{array}\right\} \tag{7.26}$$

where $\alpha(\varepsilon)$ is a continuous function of ε. If we now write the inner solution in terms of ζ and let $\varepsilon \to 0$ *with ζ fixed* then it must be the same as the outer solution written in terms of ζ under the limit $\varepsilon \to 0$ *with ζ fixed*. That is, for the intermediate region near $x = 0$, we must have

$$\lim_{\substack{\varepsilon \to 0 \\ \zeta \text{ fixed}}} \{[u_0(x) + \varepsilon u_1(x) + \ldots]_{x = \alpha(\varepsilon)\zeta}\}$$

$$= \lim_{\substack{\varepsilon \to 0 \\ \zeta \text{ fixed}}} \{[\bar{u}_0(\xi) + \varepsilon \bar{u}_1(\xi) + \ldots]_{\xi = [\alpha(\varepsilon)/\varepsilon]\zeta}\}. \tag{7.27}$$

From (7.22) in terms of ζ

$$u(x;\varepsilon) \sim U_1 e^{(1 - \alpha(\varepsilon)\zeta)} + \varepsilon U_1 (1 - \alpha(\varepsilon)\zeta) e^{(1 - \alpha(\varepsilon)\zeta)} + \ldots, \text{ as } \varepsilon \to 0$$

$$\sim U_1 e\{1 - \alpha(\varepsilon)\zeta + \ldots\} + \varepsilon U_1 (1 - \alpha(\varepsilon)\zeta) e\{1 - \alpha(\varepsilon)\zeta + \ldots\} + \ldots,$$
$$\text{as } \varepsilon \to 0$$

$$\sim U_1 e\{1 - \alpha(\varepsilon)\zeta + \ldots + \varepsilon - 2\varepsilon\alpha(\varepsilon)\zeta + \ldots\}, \text{ as } \varepsilon \to 0, \tag{7.28}$$

while from (7.23) with (7.24)

$$\bar{u}(\xi;\varepsilon) \sim \left\{\bar{a}_0 + \bar{b}_0 \exp\left(-\frac{\alpha(\varepsilon)}{\varepsilon}\zeta\right)\right\} + \varepsilon\left\{\bar{a}_1 + \bar{b}_1 \exp\left(-\frac{\alpha(\varepsilon)}{\varepsilon}\zeta\right)\right.$$

$$+ \bar{a}_0\left(1 - \frac{\alpha(\varepsilon)}{\varepsilon}\zeta\right) + \left.\bar{b}_0\frac{\alpha(\varepsilon)}{\varepsilon}\zeta \exp\left(-\frac{\alpha(\varepsilon)}{\varepsilon}\zeta\right)\right\} + \ldots$$

$$\sim \{\bar{a}_0 - \alpha(\varepsilon)\bar{a}_0\zeta\} + \varepsilon\{\bar{a}_0 + \bar{a}_1 + \ldots\} + o(\varepsilon), \text{ as } \varepsilon \to 0, \tag{7.29}$$

since $\exp\left(-\dfrac{\alpha(\varepsilon)}{\varepsilon}\zeta\right) = o(\varepsilon^N)$ for all N since $\alpha(\varepsilon)/\varepsilon \to \infty$ as $\varepsilon \to 0$.

Thus the limit process (7.27) gives, with (7.28) and (7.29), on equating corresponding terms

$$U_1 e = \bar{a}_0, \quad \bar{a}_1 + \bar{a}_0 = U_1 e, \ldots,$$

which, with (7.25), give

$$\bar{a}_0 = U_1\,\mathrm{e}, \quad \bar{b}_0 = U_0 - U_1\,\mathrm{e}, \quad \bar{a}_1 = 0, \quad \bar{b}_1 = -\bar{a}_0 = -U_1\,\mathrm{e}, \dots$$
$$(7.30)$$

Note that we do not need to determine $\alpha(\varepsilon)$ here; it simply has to behave as in (7.26). Thus the inner solution valid near $x = 0\,(\xi = 0)$, that is $0 \leq x \leq \alpha(\varepsilon)$, is, from (7.24) and (7.30), determined as

$$\bar{u}(\xi;\varepsilon) \sim \{U_1\,\mathrm{e}[1 - \mathrm{e}^{-\xi}] + U_0\,\mathrm{e}^{-\xi}\}$$
$$+\varepsilon\{U_0\xi\,\mathrm{e}^{-\xi} + U_1\,\mathrm{e}[1 - \xi - (1+\xi)\,\mathrm{e}^{-\xi}]\} + \dots, \text{ as } \varepsilon \to 0. \quad (7.31)$$

We now have a complete description of the solution namely (7.22) for $0 < x \leq 1$ and (7.31) for $0 \leq x < \alpha(\varepsilon)$. A practical question is now where and how we move from the inner solution to the outer one. We can do this by obtaining a *composite asymptotic expansion* of the solution, which is uniformly valid, by using the fact that there is an overlap domain. Such a uniformly valid asymptotic solution is necessarily more complicated than either the outer or inner solution separately. One way, the commonest in fact, to obtain a composite expansion is to add the inner solution to the outer and subtract the intermediary form so that it is not included twice. Here the intermediate solution is either the outer or inner solutions in terms of the intermediate variable ζ since they are the same. Thus

$$u_{\text{composite}} = u_{\text{inner}} + u_{\text{outer}} - u_{\text{intermediate}}. \quad (7.32)$$

With this solution, in the singular region near $x = 0$, u_{inner} is the form left while, away from this singular region, the outer form u_{outer} is left. The transition using (7.32) is thus a smooth one. With the above example the composite solution $u_c(x)$ to $O(1)$ is, using (7.31) the inner solution, (7.22) the outer solution, and (7.28) the intermediate solution,

$$u_c(x) \sim \bar{u}_0(\xi) + u_0(x) - \lim_{\substack{\varepsilon \to 0 \\ \zeta \text{ fixed}}} u_0(\alpha(\varepsilon)\zeta, \varepsilon)$$

$$= \{U_1\,\mathrm{e}[1 - \mathrm{e}^{-x/\varepsilon}] + U_0\,\mathrm{e}^{-x/\varepsilon}\} + \{U_1\,\mathrm{e}^{1-x}\} - \{U_1\,\mathrm{e}\}$$

$$= (U_0 - U_1\,\mathrm{e})\,\mathrm{e}^{-x/\varepsilon} + U_1\,\mathrm{e}^{1-x}. \quad (7.33)$$

The exact solution of the problem posed by (7.19) and (7.20) is, after some algebra,

$$u(x;\varepsilon) = (1 - \mathrm{e}^{1-(1/\varepsilon)})^{-1}\{U_1\,\mathrm{e}^{1-x}$$

$$+ (U_0 - U_1\,\mathrm{e})\,\mathrm{e}^{-x/\varepsilon} - U_0\,\mathrm{e}^{1-x}\,\mathrm{e}^{-1/\varepsilon}\},$$

which asymptotes to (7.33) for $0 < \varepsilon \ll 1$ since $e^{-1/\varepsilon} = o(\varepsilon^N)$, for all N, is negligible asymptotically.

In the above we chose the outer solution with the $u_n(x)$ as in (7.21) to satisfy the boundary condition at $x = 1$. Suppose instead we require it to satisfy the first boundary condition of (7.20), that is $u(0; \varepsilon) = U_0$. We now have for the outer solution

$$u(x; \varepsilon) \sim U_0 e^{-x} - \varepsilon U_0 x e^{-x} + \ldots, \text{ as } \varepsilon \to 0, \qquad (7.34)$$

instead of (7.22). Since this does not satisfy the boundary condition $u(1; \varepsilon) = U_1$, the second of (7.20), we would expect a singular solution valid near $x = 1$. We thus introduce, as before, $\eta = (1-x)/\varepsilon$ as the inner variable and the inner solution $\bar{u}(\eta; \varepsilon)$ is then determined by (7.19), with this transformation, namely

$$\bar{u}_{\eta\eta} - \bar{u}_\eta + \varepsilon \bar{u} = 0.$$

The terms $\bar{u}_0(\eta)$, $\bar{u}_1(\eta)$, ... in (7.23) now satisfy

$$\bar{u}_{0\eta\eta} - \bar{u}_{0\eta} = 0, \quad \bar{u}_{n\eta\eta} - \bar{u}_{n\eta} = -\bar{u}_{n-1}, \ldots, \quad n \geq 1,$$

which have *unbounded* solutions as $\eta \to \infty$. Since (7.34) is bounded as $x \to 1$ there can be no match with such an inner solution. Hence the possibility of a boundary layer at $x = 1$ does not exist and the solution is as we derived above, namely that with a singular domain at $x = 0$.

Two-point boundary value problems exist where there is no singular region at *either* boundary but can exist in the interior: O'Malley (1974) and Cohen (1977) discuss some examples of these. The books cited give very readable discussions of matching and composite expansions with worked examples.

Exercises

1. Use singular perturbation concepts to find accurate approximations for three solutions of $\varepsilon x^3 - x + 1 = 0$ as $\varepsilon \downarrow 0$.

2. Use a matched asymptotic procedure, as $\varepsilon \downarrow 0$, to find uniformly valid solutions of the following and compare your results with the exact solution:
   (i) $\varepsilon u' + u = 1$, $u(0) = 0$.
   (ii) $\varepsilon u'' - u = -1$, $u(-1) = 0 = u(1)$. (The stretching involves $\varepsilon^{1/2}$ in this problem.)
   (iii) $\varepsilon u'' + u' = a + bx$, a, b constants, $u(0) = 0$, $u(1) = 1$.

3. Consider the problem

$$\varepsilon u'' + a(x)u' + b(x)u = 0, \quad u(0) = A, \quad u(1) = B,$$

where $a(x) > 0$ in $0 \leq x \leq 1$ and $b(x)$ is bounded. Show that as $\varepsilon \downarrow 0$ a uniformly valid (composite) asymptotic solution to $O(1)$ is

$$u(x; \varepsilon) \sim u_0(x) + [A - u_0(0)] \exp[-a(0)x/\varepsilon],$$

where $u_0(x) = B \exp\left(\int_x^1 \frac{b(s)}{a(s)} ds\right).$

7.2.   Method of multiple scales and suppression of secular terms

Again we shall introduce the concepts and develop the method by way of a specific problem.

Consider the linear equation for a simple damped harmonic oscillator, namely,

$$\frac{d^2u}{dt^2} + \varepsilon \frac{du}{dt} + u = 0, \quad 0 < \varepsilon \ll 1, \tag{7.35}$$

where we associate t with time. The damping comes from the ε-term in the equation. The general solution of this equation is

$$u(t; \varepsilon) = A e^{-\varepsilon t/2} \cos\left(\sqrt{\left(1 - \frac{\varepsilon^2}{4}\right)}t + B\right), \tag{7.36}$$

where A and B are arbitrary constants. The effect of the damping term is to alter the frequency from 2π, the undamped frequency, to $2\pi/\sqrt{(1 - \varepsilon^2/4)}$ which for ε small is approximately 2π. However the amplitude is slowly decreased and as $t \to \infty$, $u \to 0$ for any $\varepsilon > 0$, however small. If $\varepsilon = 0$, u simply oscillates between $\pm A$ for all time. Thus as in §7.1 the solution with $\varepsilon = 0$ differs fundamentally from that with a positive but very small ε: the problem is thus singular. However it does so only after a long time when $e^{-\varepsilon t}$ is not approximately 1, in other words when εt is $O(1)$, that is for t large $O(1/\varepsilon)$. For times $t = O(1)$ the solution (7.36) is approximately the same as the solution with $\varepsilon = 0$. Thus there are in effect two time scales associated with this problem, that in which the time is $O(1)$, the fast time say, and that in which t is large like $O(1/\varepsilon)$, the slow time say during which the amplitude is appreciably decreased. By expanding $\sqrt{(1-\varepsilon^2/4)}t$ as $(1-\frac{1}{8}\varepsilon^2 + O(\varepsilon^4))t$ we see that there are other, even longer, time scales associated with $\varepsilon^2 t$, $\varepsilon^4 t$ and so on: these affect the frequency. The method of multiple scales is based on the concept of different time scales.

Suppose we look, naively again, for a regular perturbation solution to (7.35) in the form

$$u(t; \varepsilon) = \sum_{n=0}^{\infty} \varepsilon^n u_n(t), \qquad (7.37)$$

since $0 < \varepsilon \ll 1$. Substitution in (7.35) and equating powers of ε gives

$$\frac{d^2 u_0}{dt^2} + u_0 = 0, \quad \frac{d^2 u_n}{dt^2} + u_n = -\frac{du_{n-1}}{dt}, \quad n \geq 1. \qquad (7.38)$$

By way of example let us take as initial conditions for (7.35)

$$u(0; \varepsilon) = a, \quad \frac{du}{dt}\bigg|_{t=0} = 0, \qquad (7.39)$$

which from (7.37), require the $u_n(t)$ to satisfy,

$$u_0(0) = a, \quad u_n(0) = 0, \quad n \geq 1, \quad \frac{du_n}{dt}\bigg|_{t=0} = 0, \quad n \geq 0.$$

Solutions of (7.38) are thus

$$u_0(t) = a \cos t, \quad u_1(t) = \tfrac{1}{2}a \sin t - \tfrac{1}{2}at \cos t, \ldots$$

and so (7.37) becomes

$$u(t; \varepsilon) \sim a \cos t + \tfrac{1}{2}\varepsilon a(\sin t - t \cos t) + \ldots, \quad \varepsilon \to 0. \qquad (7.40)$$

This solution is not a uniformly valid asymptotic expansion for all t since $\varepsilon u_1(t) \neq o(u_0(t))$ for t large because of the $\varepsilon t \cos t$ term in $\varepsilon u_1(t)$. Clearly higher-order contributions in (7.37) will involve terms like $t^n \cos t$ and $t^n \sin t$. Such nonperiodic terms are called *secular terms*. As long as $t = O(1)$, the series solution (7.40) suffices and is convergent for ε small enough. It ceases to be so when $t = O(1/\varepsilon)$ and so it is not a uniformly valid solution for all time. For times $O(1)$ we expect (7.40) to be a good approximation. The exact solution (7.36) with initial conditions (7.39) is

$$u(t;\varepsilon) = a\,e^{-\varepsilon t/2}\left[\cos\sqrt{\left(1-\frac{\varepsilon^2}{4}\right)}t + \frac{\varepsilon}{2\sqrt{\left(1-\frac{\varepsilon^2}{4}\right)}}\sin\sqrt{\left(1-\frac{\varepsilon^2}{4}\right)}t\right].$$

$$(7.41)$$

If we expand (7.41) as a Taylor series for $0 < \varepsilon \ll 1$ and $t = O(1)$ we get (7.40) as we expected. It is however the expansion of $e^{-\varepsilon t/2}$ which gives rise to the secular terms. Finally, note from the solution that the time involved in the oscillatory part appears as $\sqrt{(1-\varepsilon^2/4)}t$ which we associate with the fast time whereas the εt in $e^{-\varepsilon t/2}$ is associated with the slow time: we denote these times respectively by t^* and τ. In this example then we have

$$t^* = \sqrt{\left(1-\frac{\varepsilon^2}{4}\right)}t = t(1-\tfrac{1}{8}\varepsilon^2+O(\varepsilon^4)), \quad \tau = \varepsilon t. \qquad (7.42)$$

The method of multiple scales, or the two time method as it is often referred to in this type of problem, considers the solution $u(t;\varepsilon)$ of (7.35) as a function of two time variables, a fast one t^* and a slow one τ. We then look for an asymptotic solution in the form

$$u(t;\varepsilon) \sim u_0(t^*,\tau)+\varepsilon u_1(t^*,\tau)+\ldots, \qquad (7.43)$$

with

$$\tau = \varepsilon t, \quad t^* = t(1+w_1\varepsilon^2+w_2\varepsilon^3+\ldots), \qquad (7.44)$$

where the constants $w_1, w_2, \ldots$ are to be determined. The reason for leaving out the $O(\varepsilon)$ term in the transformation for t^* is that it would give an εt in t^*: this we have defined separately as τ and we do not wish to have t^* dependent explicitly on τ. With these transformations (7.44)

$$\left.\begin{aligned}
\frac{d}{dt} &= (1+w_1\varepsilon^2+\ldots)\frac{\partial}{\partial t^*}+\varepsilon\frac{\partial}{\partial\tau}, \\
\frac{d^2}{dt^2} &= (1+2w_1\varepsilon^2+\ldots)\frac{\partial^2}{\partial t^{*2}}+2\varepsilon(1+w_1\varepsilon^2+\ldots)\frac{\partial^2}{\partial t^*\partial\tau}+\varepsilon^2\frac{\partial^2}{\partial\tau^2}.
\end{aligned}\right\}$$

$$(7.45)$$

We now substitute (7.43) with (7.44) and (7.45) into the differential equation (7.35) and equate corresponding powers of ε in the usual way to get, using the same initial conditions (7.39), the series of *partial* differential equations.

$$
\left.
\begin{aligned}
&\frac{\partial^2 u_0}{\partial t^{*2}} + u_0 = 0, \quad u_0(0, 0) = a, \quad \frac{\partial u_0}{\partial t^*}(0, 0) = 0, \\[2mm]
&\frac{\partial^2 u_1}{\partial t^{*2}} + u_1 = -\frac{\partial u_0}{\partial t^*} - 2\frac{\partial^2 u_0}{\partial t^* \, \partial \tau}, \\[2mm]
&u_1(0, 0) = 0, \quad \frac{\partial u_1}{\partial t^*}(0, 0) + \frac{\partial u_0}{\partial \tau}(0, 0) = 0, \\[2mm]
&\frac{\partial^2 u_2}{\partial t^{*2}} + u_2 = -\frac{\partial u_1}{\partial t^*} - \frac{\partial u_0}{\partial \tau} - 2w_1 \frac{\partial^2 u_0}{\partial t^{*2}} - \frac{\partial^2 u_0}{\partial \tau^2} - 2\frac{\partial^2 u_1}{\partial t^* \partial \tau}, \\[2mm]
&u_2(0, 0) = 0, \quad \frac{\partial u_2}{\partial t^*}(0, 0) + \frac{\partial u_1}{\partial \tau}(0, 0) + w_1 \frac{\partial u_0}{\partial t^*}(0, 0) = 0,
\end{aligned}
\right\} \quad (7.46)
$$

. . . .

Although it might appear that the problem is now more complicated since we have to solve a series of partial differential equations, they are in fact essentially ordinary differential equations since the left-hand side operator only involves t^* and so τ is in effect a parameter.

The solution of the first of (7.46) is

$$u_0(t^*, \tau) = A_0(\tau) \cos t^* + B_0(\tau) \sin t^*,$$

$$A_0(0) = a, \quad B_0(0) = 0.$$

Note that the A_0 and B_0 are undetermined functions of τ at this stage. Using this solution in the u_1-equation in (7.46) we have

$$
\left.
\begin{aligned}
&\frac{\partial^2 u_1}{\partial t^{*2}} + u_1 = \left(2\frac{dA_0}{d\tau} + A_0\right) \sin t^* - \left(2\frac{dB_0}{d\tau} + B_0\right) \cos t^*, \\[2mm]
&u_1(0, 0) = 0, \quad \frac{\partial u_1}{\partial t^*}(0, 0) = -\frac{dA_0}{d\tau}(0).
\end{aligned}
\right\} \quad (7.47)
$$

The particular solution for $u_1(t^*, \tau)$ will involve secular terms $t^* \cos t^*$, $t^* \sin t^*$ unless the coefficients of $\sin t^*$ and $\cos t^*$ are both zero. We saw in the regular perturbation solution (7.40) that the appearance of such terms restricted the domain of validity of the solution to times $O(1)$. Since $A_0(\tau)$, $B_0(\tau)$ are still undetermined at this stage we can choose them in such a way that secular terms in $u_1(t^*, \tau)$ in (7.47) are suppressed. This is effected if the coefficients of $\sin t^*$ and $\cos t^*$ are zero, that is

$$2\frac{dA_0}{d\tau} + A_0 = 0, \quad 2\frac{dB_0}{d\tau} + B_0 = 0 \tag{7.48}$$

the solutions of which are, using the initial conditions required by (7.46)

$$A_0(\tau) = a\,e^{-\tau/2}, \quad B_0(\tau) = 0. \tag{7.49}$$

At this stage $u_0(t^*, \tau)$ is completely determined as

$$u_0(t^*, \tau) = a\,e^{-\tau/2}\cos t^*$$

and so the $O(1)$ asymptotic solution as $\varepsilon \to 0$ is

$$u(t; \varepsilon) = u(t^*, \tau; \varepsilon) \sim a\,e^{-\varepsilon t/2}\cos t + \ldots,$$

Note that to $O(1)$ we have $t^* \sim t$. This is the $O(1)$ uniformly valid asymptotic solution *for all* $t \geq 0$ and agrees with the $O(1)$ term from the exact solution (7.41).

We now solve (7.47) for $u_1(t^*, \tau)$, with (7.48) to get

$$\left.\begin{array}{l} u_1(t^*, \tau) = A_1(\tau)\cos t^* + B_1(\tau)\sin t^*, \\[4pt] A_1(0) = 0, \quad B_1(0) = a/2. \end{array}\right\} \tag{7.50}$$

The $A_1(\tau)$, $B_1(\tau)$ are determined, in a similar way to $A_0(\tau)$, $B_0(\tau)$, by choosing them to suppress secular terms in the next order equation, the one for $u_2(t^*, \tau)$ which is from (7.46), using (7.49) for $A_0(\tau)$,

$$\frac{\partial^2 u_2}{\partial t^{*2}} + u_2 = \left(2\frac{dA_1}{d\tau} + A_1\right)\sin t^*$$

$$- \left(2\frac{dB_1}{d\tau} + B_1 + \frac{d^2 A_0}{d\tau^2} + \frac{dA_0}{d\tau} - 2w_1 A_0\right)\cos t^*.$$

To suppress secular terms at this stage we require the coefficients of $\sin t^*$ and $\cos t^*$ to be zero, that is $A_1(\tau)$, $B_1(\tau)$ satisfy

$$\left.\begin{array}{l} 2\dfrac{dA_1}{d\tau} + A_1 = 0, \quad 2\dfrac{dB_1}{d\tau} + B_1 = 2w_1 A_0 - \dfrac{dA_0}{d\tau} - \dfrac{d^2 A_0}{d\tau^2} \\[10pt] \hspace{4cm} = 2a(w_1 + \tfrac{1}{8})\,e^{-\tau/2}, \\[8pt] A_1(0) = 0, \quad B_1(0) = a/2. \end{array}\right\} \tag{7.51}$$

Clearly $A_1(\tau) \equiv 0$. The solution for $B_1(\tau)$ is

$$B_1(\tau) = \tfrac{1}{2}a\,e^{-\tau/2} + a(w_1 + \tfrac{1}{8})\tau\,e^{-\tau/2},$$

which gives $u(t;s)$ to $O(\varepsilon)$ as

$$u(t;\varepsilon) = u(t^*,\tau;\varepsilon) \sim a\,e^{-\tau/2}\{\cos t^* + \varepsilon[\tfrac{1}{2} + (w_1 + \tfrac{1}{8})\tau]\sin t^* + \ldots\}.$$

The appearance of the $\varepsilon\tau$ term gives rise to another kind of non-uniformity, this time as $\varepsilon\tau$ becomes large which, in real time is for t large, $O(1/\varepsilon^2)$ for example. But w_1 is still at our disposal and so if we choose $w_1 = -\tfrac{1}{8}$ the term $\tau\varepsilon^{-\tau/2}$ does not appear in $B_1(\tau)$. This we do and so $B_1(\tau) = \tfrac{1}{2}a\,e^{-\tau/2}$ and we now have a uniformly valid asymptotic solution to $u(t;\varepsilon)$ as

$$u(t;\varepsilon) = u(t^*,\tau;\varepsilon) \sim a\,e^{-\tau/2}\{\cos t^* + \tfrac{1}{2}\varepsilon\sin t^* + \ldots\},$$

$$\tau = \varepsilon t, \quad t^* = t[1 - \tfrac{1}{8}\varepsilon^2 + \ldots]. \quad\quad\quad\left.\right\} \quad (7.52)$$

In terms of t this gives to $O(\varepsilon)$

$$u(t;\varepsilon) \sim a\,e^{-\varepsilon t/2}\{\cos t(1 - \tfrac{1}{8}\varepsilon^2 + \ldots) + \frac{\varepsilon}{2}\sin t(1 - \tfrac{1}{8}\varepsilon^2 + \ldots) + \ldots\},$$

which is the same as that given by the exact solution (7.41) on expanding $\sqrt{(1 - \varepsilon^2/4)}$ and retaining the comparably ordered terms.

In summary then, the essentials of this method are to introduce two time scales, a fast one t^* and a slow one τ and then seek a regular perturbation solution for $u(t^*,\tau;\varepsilon)$ as $\varepsilon \to 0$ by systematically suppressing any terms which would give rise to *secular behaviour and nonuniformity* in either of the time variables. There are other versions of this method but they are all similar in principle in that they choose unknown parameters or functions to suppress secular terms. One of them, for example, has a single transformation of the time, say

$$t = \bar{t} + \varepsilon f_1(\bar{t}) + \ldots.$$

Then writing $u(t;\varepsilon)$ as $u(\bar{t};\varepsilon)$ a regular solution is sought in the form

$$u(\bar{t};\varepsilon) \sim u_0(\bar{t}) + \varepsilon u_1(\bar{t}) + \ldots,$$

and the $f_1(\bar{t}), f_2(\bar{t}), \ldots$ chosen to suppress secular terms: this method will be required in one of the exercises.

We consider now a nonlinear problem for which we do not have an exact analytical solution, namely the van der Pol equation

$$\frac{d^2 u}{dt^2} + u = \varepsilon(1 - u^2)\frac{du}{dt}, \quad \varepsilon > 0. \quad\quad (7.53)$$

This equation has a periodic limit cycle solution, that is a periodic solution of fixed amplitude which is independent of the initial conditions. We shall obtain the uniformly valid asymptotic solution to $O(1)$ as $\varepsilon \to 0$ using the multiple scale method of suppression of secular terms.

As before we introduce fast (t^*) and slow (τ) time scales, using (7.44) and seek a solution for u as a function of t^* and τ in the form of a regular perturbation solution as in (7.43). Let us choose the initial conditions as

$$u(0; \varepsilon) = a, \quad \frac{du}{dt}(0; \varepsilon) = b, \tag{7.54}$$

where a, b are given constants independent of ε. The first two equations in the equivalent of (7.46) are

$$\left. \begin{array}{l} \dfrac{\partial^2 u_0}{\partial t^{*2}} + u_0 = 0, \quad u_0(0, 0) = a, \quad \dfrac{\partial u_0}{\partial t^*}(0, 0) = b, \\[2mm] \dfrac{\partial^2 u_1}{\partial t^{*2}} + u_1 = -2\dfrac{\partial^2 u_0}{\partial t^* \partial \tau} + (1 - u_0^2)\dfrac{\partial u_0}{\partial t^*}, \\[2mm] u_1(0, 0) = 0, \quad \dfrac{\partial u_1}{\partial t^*}(0, 0) + \dfrac{\partial u_0}{\partial \tau}(0, 0) = 0. \end{array} \right\}$$

Thus, with (7.54),

$$\left. \begin{array}{l} u_0(t^*, \tau) = A_0(\tau) \cos t^* + B_0(\tau) \sin t^*, \\[2mm] A_0(0) = a, \quad B_0(0) = b, \end{array} \right\} \tag{7.55}$$

and the equation for u_1 becomes

$$\begin{aligned} \frac{\partial^2 u_1}{\partial t^{*2}} + u_1 &= -2\left[-\frac{dA_0}{d\tau} \sin t^* + \frac{dB_0}{d\tau} \cos t^* \right] \\ &\quad + [1 - A_0^2 \cos^2 t^* - B_0^2 \sin^2 t^* - 2A_0 B_0 \cos t^* \sin t^*] \\ &\quad \times [-A_0 \sin t^* + B_0 \cos t^*] \\ &= \left[2\frac{dA_0}{d\tau} - A_0 + \tfrac{1}{4}A_0(A_0^2 + B_0^2) \right] \sin t^* \\ &\quad - \left[2\frac{dB_0}{d\tau} - B_0 + \tfrac{1}{4}B_0(A_0^2 + B_0^2) \right] \cos t^* \\ &\quad + \tfrac{1}{4}A_0[A_0^2 - 3B_0^2] \sin 3t^* + \tfrac{1}{4}B_0[B_0^2 - 3A_0^2] \cos 3t^*. \end{aligned}$$

The sin t^*, cos t^* terms on the right are the ones which give rise to secular terms. So, to suppress secular terms in $u_1(t^*, \tau)$, $A_0(\tau)$, $B_0(\tau)$ must satisfy, with initial conditions (7.55),

$$2\frac{\mathrm{d}A_0}{\mathrm{d}\tau} = A_0[1 - \tfrac{1}{4}(A_0^2 + B_0^2)], \quad 2\frac{\mathrm{d}B_0}{\mathrm{d}\tau} = B_0[1 - \tfrac{1}{4}(A_0^2 + B_0^2)],$$

$$A_0(0) = a, \quad B_0(0) = b. \tag{7.56}$$

These are less trivial to solve than in the last example. However the symmetry of these equations suggests multiplying the first by A_0, the second by B_0 and adding to get

$$\frac{\mathrm{d}}{\mathrm{d}\tau}(A_0^2 + B_0^2) = (A_0^2 + B_0^2)[1 - \tfrac{1}{4}(A_0^2 + B_0^2)]$$

and so, after some algebra

$$A_0^2(\tau) + B_0^2(\tau) = \frac{C\,\mathrm{e}^\tau}{1 + \tfrac{1}{4}C\,\mathrm{e}^\tau}, \quad C = \frac{4(a^2 + b^2)}{4 - (a^2 + b^2)}. \tag{7.57}$$

If we now multiply the first of (7.56) by A_0 it becomes

$$\frac{1}{A_0^2}\frac{\mathrm{d}A_0^2}{\mathrm{d}\tau} = 1 - \frac{C\,\mathrm{e}^\tau}{4 + C\,\mathrm{e}^\tau}$$

and so with $A_0(0) = a$, and (7.57)

$$A_0(\tau) = a\left[\frac{1 - d}{1 - d\,\mathrm{e}^{-\tau}}\right]^{1/2}, \quad d = 1 - \frac{4}{a^2 + b^2},$$

$$B_0(\tau) = b\left[\frac{1 - d}{1 - d\,\mathrm{e}^{-\tau}}\right]^{1/2}.$$

The uniformly valid $O(1)$ asymptotic approximation as $\varepsilon \to 0$ to the limit cycle solution of the van der Pol equation (7.53) is thus

$$u(t;\varepsilon) \sim \left[\frac{1 - d}{1 - d\,\mathrm{e}^{-\varepsilon t}}\right]^{1/2}(a\cos t + b\sin t), \quad 0 < \varepsilon \ll 1,$$

$$= \left[\frac{4/(a^2 + b^2)}{1 - d\,\mathrm{e}^{-\varepsilon t}}\right]^{1/2}(a^2 + b^2)^{1/2}\sin(t + \theta), \quad 0 = \tan^{-1}(b/a).$$

Thus

$$u(t;\varepsilon) \sim \frac{2}{[1 - d\,\mathrm{e}^{-\varepsilon t}]^{1/2}}\sin(t + \theta), \quad \theta < \varepsilon \ll 1, \tag{7.58}$$

with $d = 1 - 4/(a^2 + b^2)$ and $\theta = \tan^{-1}(b/a)$.

As $t \to \infty$, $u(t; \varepsilon) \sim 2 \sin(t + \theta)$ which is periodic with period 2π and amplitude 2. Thus irrespective of the initial conditions, that is a and b, the *amplitude* of the periodic solution is 2: this is a classic example of a limit cycle.

Methods of suppression of secular terms are very general and widely applicable to equations of the type

$$\frac{d^2u}{dt^2} + \omega^2 u = \varepsilon f\left(u, \frac{du}{dt}, \varepsilon\right), \qquad (7.59)$$

where ω is a constant and f a *nonlinear* function of u, du/dt and ε. There is a variety of such methods, for example, Poincaré's method of suppression of secular terms, the method of averages, the method of Bogliubov and Metropolsky and so on. The books by Kevorkian and Cole (1981) and Nayfeh (1973) have numerous examples of such methods. Those in which ω^2 may depend on εt and f is linear in u and du/dt were considered in §6.2, using the exponential method: for linear problems it is somewhat easier and less complicated than the method of suppression of secular terms.

Exercises

1. Find the first two terms in the uniformly valid asymptotic solution, as $\varepsilon \to 0$, of

$$\frac{d^2u}{dt^2} + \varepsilon \frac{du}{dx} + u = 0, \quad u(0) = 0, \quad \frac{du}{dt}(0) = b.$$

2. Find the first term in a uniformly valid asymptotic solution, as positive $\varepsilon \to 0$, of

$$\frac{d^2u}{dt^2} + u = \varepsilon(u^2 - 1)\frac{du}{dt}, \quad u(0; \varepsilon) = a, \quad \frac{du}{dt}(0; \varepsilon) = 0.$$

3. Consider the problem of finding the solution of $d^2u/dt^2 + \omega^2 u = \varepsilon u^3$ where ω is a constant and $0 < \varepsilon \ll 1$ when $u(0; \varepsilon) = a$, $(du/dt)(0; \varepsilon) = 0$, where a is independent of ε. Introduce a new time scale $\bar{t}$ by writing

$$t = \bar{t} + \varepsilon f_1(\bar{t}) + \dots$$

where $f_1, f_2, \dots$ are undetermined functions of $\bar{t}$.

Express the problem in terms of $\bar{t}$ and look for a solution in the form

$$u = \sum_{n=0}^{\infty} \varepsilon^n u_n(\bar{t})$$

by suppressing secular terms. Hence show that $f_1(\bar{t}) = (3a^2/8\omega^2)\,\bar{t}$ and that

$$u(t; \varepsilon) \sim a \cos \omega (1 - \varepsilon \frac{3a^2}{8\omega^2} + \ldots)t \text{ as } \varepsilon \to 0.$$

4. Find the first term in the uniformly valid asymptotic approximation to the limit cycle solution of the damped van der Pol equation

$$\frac{d^2u}{dt^2} + u = \varepsilon(1 - u^2)\frac{du}{dt} - \varepsilon u^3, \quad 0 < \varepsilon \ll 1.$$

5. Show that the first term in a uniformly valid asyptotic approximation to the solution of

$$\frac{d^2u}{dt^2} + u = -\varepsilon \left[\left(\frac{du}{dt}\right)^3 + \alpha \frac{du}{dt} \right], \quad 0 < \varepsilon \ll 1,$$

with $u(0; \varepsilon; \alpha) = 0$, $(du/dt)(0; \varepsilon; \alpha) = a$, where a and α are positive constants independent of ε is given by

$$u(t; \varepsilon) \sim \left[1 - \frac{3a^2 e^{-\alpha \varepsilon t}}{3a^2 + 4\alpha} \right] \left[\frac{4\alpha a^2}{3a^2 + 4\alpha} \right]^{1/2} e^{-\alpha \varepsilon t/2} \sin t.$$

6. Find the first term in the uniformly valid asymptotic solution of

$$\frac{d^2u}{dt^2} + u = -\varepsilon u^3, \quad u(0; \varepsilon) = a, \quad \frac{du}{dt}(0; \varepsilon) = 0, \quad 0 < \varepsilon \ll 1.$$

Bibliography

Carrier, G. F., Krook, M. and Pearson, C. E. (1966). *Functions of a complex variable*. McGraw-Hill, New York.

Cohen, D. S. (1977). *Perturbation theory* in Lectures in Applied Mathematics (Vol. 16) on *Modern modelling of continuum phenomena* (Editor R. C. Di Prima). American Mathematical Society, Providence.

Cole, J. D. (1968). *Perturbation methods in applied mathematics*. Blaisdell, Waltham, Mass.

Copson, E. T. (1965). *Asymptotic expansions*. Cambridge University Press, Cambridge.

de Bruijn, N. G. (1958). *Asymptotic methods in analysis*. North-Holland Publishing Co., Amsterdam.

Dingle, R. B. (1973). *Asymptotic expansions: their derivation and interpretation*. Academic Press, New York.

Erdelyi, A. (1956). *Asymptotic expansions*. Dover Publications Inc., New York.

Jeffreys, H. (1962). *Asymptotic approximations*. Clarendon Press, Oxford.

Kevorkian, J. and Cole, J. D. (1981). *Perturbation methods in applied mathematics*. Springer-Verlag, New York.

Murray, J. D. (1977). *Nonlinear differential equation models in biology*. Clarendon Press, Oxford.

Nafeh, A. H. (1973). *Perturbation methods*. Wiley, New York.

Olver, F. W. J. (1975). *Asymptotic and special functions*. Academic Press, New York.

O'Malley, R. E. (1974). *Introduction to singular perturbations*. Academic Press, New York.

Van Dyke, M. (1975). *Perturbation methods in fluid mechanics*. Panabolic Press, Stanford.

Wasow, W. (1965). *Asymptotic expansions for ordinary differential equations*. Interscience Publishers, New York.

Watson, G. N. (1952). *Theory of Bessel functions* (2nd edn.). Cambridge University Press, Cambridge.

Index

Airy, G. B., 54
  equation, 54, 70, 115, 133
  functions, 54, 133
  function asymptotic expansions, 60, 61, 69, 70, 134
  integral, 57
Asymptotic:
  equivalence, 4
  expansion, 5
  expansion (definition), 12
  expansion (uniqueness), 12
  matching process, 133
  power series, 13
  sequence, 11

Bessel:
  equation, 55, 105, 133
  functions, 61
  function asymptotic forms, 65, 78, 105, 137
Bleistein, N., 51
de Bruijn, N. G., 16, 161
Burkill, J. C., 101

Carrier, G. F., 51, 161
Cauchy–Riemann equations, 43
Chester, C., 51
Cohen, D. S., 138, 150, 161
Cole, J. D., 2, 138, 159, 161
Composite expansion, 149
Contour integral solutions, 55
Convergent power series, 4
Copson, E. T., 19, 161

Debye, P., 40
Differential equations, 99
  asymptotic solutions involving a parameter, 111
Differentiation of asymptotic series, 16
Dingle, R. B., 161
Dispersion relation, 80
Divergent series, 8
Dominant term, 13

Erdelyi, A., 99, 161
Error function, 20, 27
Exponential function:
  complex argument, 9
  real argument, 6, 20

Fourier transform, 86
  asymptotic expansions, 90, 91
Friedman, B., 51
Frobenius method, 101

Gamma function:
  analytic continuation, 54
  complex argument, 51
  real argument, 19, 37
Garabedian, P. R., 127
Gauge function, 3
Green, G., 111
Group velocity, 84

Handelsman, R. A., 51
Hankel function, 62
  asymptotic forms, 65
Hankel integral, 68
Horn, J., 111

Ince, E. L., 101
Inner solution, 145
Integration:
  by parts method, 19
  of asymptotic series, 15
Intermediate:
  region, 141
  variable, 148

Jeffreys, B. S., 81
Jeffreys, H., 51, 81, 161

Kelvin, Lord, 72
Kevorkian, J., 138, 159, 161
Krook, M., 51, 161

Lamb, H., 84
de Laplace, P. S., 29
   integral, 21
   method, 28
Laplace transform, 86, 96
   asymptotic expansion, 97
Lighthill, M. J., 81
Limit cycle, 158
Liouville, J., 111
Liouville equation, 111
   asymptotic solutions, 114

Matched asymptotic procedure, 118
Murray, J. D., 138, 161

Nayfeh, A. H., 138, 159, 161
   nonlinear, 138, 159
   nonsingular solution, 138, 159

Ockendon, J. R., 121
O'Malley, R. E., 138, 159, 161
O-order, 3
o-order, 3
Order notation, 2
Outer solution, 145

Parabolic cylinder functions, 108
   asymptotic forms, 109
Pearson, C., 51, 161
Poincaré, H., 2, 12, 159
Power series, 5
   integration, 15
   differentiation, 16

Regular perturbation, 141
Riemann, B., 40

Saddle point, 43
   method, 40

Schrödinger equation, 127
   short wave length solutions, 127
   spherically symmetric form, 131
Secular terms, 121, 152
Singular:
   perturbation theory, 99, 122, 138
   region, 141
   solution, 145
Singularity, 100
   regular, 100
   irregular, 101
Sneddon, I. N., 86
Stationary phase, 72
Steepest descents, 40
Stirling's formula, 38
Stokes, G. G., 72
Stretching transformation, 143

Transition point, 128, 131
Two time method, 153

Ursell, F., 51

Van der Pol equation, 156, 160
Van Dyke, M., 2, 138

Wasow, W., 99, 161
Watson, G. N., 16, 21, 161
   lemma, 19
Wave:
   frequency, 80
   group velocity, 84
   high frequency oscillations, 125
   motion, 79
   number, 80
   slowly varying wave train, 82
   speed, 80
Whitham, G. B., 84
Whittaker, E. T., 16
WKB method, 111

Applied Mathematical Sciences

36. Bengtsson/Ghil/Källén: **Dynamic Meterology: Data Assimilation Methods.**
37. Saperstone: **Semidynamical Systems in Infinite Dimensional Spaces.**
38. Lichtenberg/Lieberman: **Regular and Stochastic Motion.** (cloth)
39. Piccinini/Stampacchia/Vidossich: **Ordinary Differential Equations in R^n.**
40. Naylor/Sell: **Linear Operator Theory in Engineering and Science.** (cloth)
41. Sparrow: **The Lorenz Equations: Bifurcations, Chaos, and Strange Attractors.**
42. Guckenheimer/Holmes: **Nonlinear Oscillations, Dynamical Systems and Bifurcations of Vector Fields.**
43. Ockendon/Tayler: **Inviscid Fluid Flows.**
44. Pazy: **Semigroups of Linear Operators and Applications to Partial Differential Equations.**
45. Glashoff/Gustafson: **Linear Optimization and Approximation: An Introduction to the Theoretical Analysis and Numerical Treatment of Semi-Infinite Programs.**
46. Wilcox: **Scattering Theory for Diffraction Gratings.**
47. Hale et al.: **An Introduction to Infinite Dimensional Dynamical Systems – Geometric Theory.**
48. Murray: **Asymptotic Analysis.**